戎珵璐 主 编

实用服装专业英语

PRACTICAL FASHION ENGLISH

東華大學出版社
·上海·

图书在版编目(CIP)数据

实用服装专业英语/戎理璐 主编. —上海：东华大学出版社，2022.10

ISBN 978-7-5669-2100-0

Ⅰ. ①实… Ⅱ. ①戎… Ⅲ. ①服装—英语—教材 Ⅳ. ①TS941

中国版本图书馆CIP数据核字(2022)第142256号

实用服装专业英语

Practical Fashion English

戎理璐 主编

责任编辑 曹晓虹

封面设计 书研社

出版发行 东华大学出版社(上海市延安西路1882号 邮政编码:200051)

联系电话 编辑部 021-62379902

营销中心 021-62193056 62373056

天猫旗舰店 http://dhdx.tmall.com

出版社网址 http://dhupress.dhu.edu.cn

印　刷 上海当纳利印刷有限公司

开　本 710mm×1000mm 1/16　印　张 6.25　字　数 148千字

版　次 2022年10月第1版　印　次 2022年10月第1次

书号:ISBN 978-7-5669-2100-0　定价:78.00元

前 言

当今的世界，相互依存、互相促进才是可持续发展之道。我们中国的学生要主动拥抱机遇，能沟通，能交流。把中国文化传扬到世界的各个角落是我们每一位中国学生的责任和使命。因此，跨过语言关，是我们完成宏伟使命的其中一步。

《实用服装专业英语》作为服装专业的配套教材，突出了专业特点，满足了岗位基本要求和为专业服务的目的，根据学校人才培养方案及设计师实践经验编写而成，针对性较强。

在教材设计中，《实用服装专业英语》的 Section A 部分，结合当今欧美国家年轻人真实口语表达并借鉴热门社交平台中博主的日常口语训练建议，突出任务的实用性以及真实性，能帮助服装专业的学生提高日常口语表达水平。教材中 Section B 部分的一些图片是由达利（中国）集团的设计师们结合工作中的真实案例挑选而出，能帮助服装专业的学生掌握在服装设计师以及设计助理岗位真实工作时所需的英语知识及技能。

本教材内容丰富，通俗易懂，形式活泼，图文并茂，可作为高职高专服装类专业学生的学习用书，也可作为中等职业院校、职业培训的教材，还可作为服装行业从业人员和服装爱好者的参考用书。

本书在编写过程中得到达利女装学院和达利（中国）集团领导的大力支持和帮助，在此深表感谢。由于时间所限，书中难免存在不妥之处，敬请同行专家和读者批评指正。

戎理璐

2022 年 7 月

目 录

Project One Fashion Trend Forecasting 流行趋势预测 ······ *1*

Module One —— Color forecasting 色彩预测 //*1*

Module Two —— Apparel forecasting Ⅰ 服装预测Ⅰ //*8*

Module Three —— Apparel forecasting Ⅱ 服装预测Ⅱ //*14*

Module Four —— Graphic & print 图案和印花 //*21*

Project Two Fashion Design and Styling 服装设计与造型配饰 ······ *28*

Module One —— Silhouette & Details Ⅰ 廓形和细节Ⅰ //*28*

Module Two —— Silhouette & Details Ⅱ 廓形和细节Ⅱ //*38*

Module Three —— Draping 立体裁剪 //*47*

Module Four —— Fabric 面料 //*53*

Module Five —— Accessory 服装配饰 //*60*

Project Three Marketing 服装营销 ······ *65*

Module One —— Sales Ⅰ 服装销售Ⅰ //*65*

Module Two —— Sales Ⅱ 服装销售Ⅱ //*70*

Module Three —— Sales Ⅲ 服装销售Ⅲ //*77*

Module Four —— After-sale service 售后服务 //*84*

Module Five —— Shopping 购物 //*88*

Reference(参考文献) ······ *93*

Project One　Fashion Trend Forecasting
流行趋势预测

Module One —— Color forecasting　色彩预测

Section A: Speaking

What is your favorite color?

Mike: Hey, Nancy. What's up?

Nancy: Hey, Mike. What's up?

Mike: I have a question for you. What's your favorite color?

Nancy: Favorite color? Green.

Mike: Yeah. Why green?

Nancy: I don't know. These days I kind of like green, but it changes all the time.

Mike: Like seasonally? You have your favorite color for the season?

Nancy: I think it's more of a monthly thing.

Mike: Right.

Nancy: You?

Mike: Well, purple. I love the color purple.

Nancy: Why is that?

Mike: I don't know really. Same thing I guess, and I never wear purple. I mean, I don't have purple clothes. I don't have anything purple in my room.

Nancy: You should have a purple suit.

Mike: A zoot suit. A purple suit.

Nancy: Yeah, just for one on those days when you're feeling it.

Mike: Yeah, I think it will work.

Nancy: OK, which color you don't like?

Mike: A color I don't like? I'll show you a color I don't like, but I don't know if there's a name for it. The closest name would probably be mustard.

Nancy: Like this green.

Mike: You think that's a green. See, I think that's more of baby diarrhea.

Nancy: It does actually look like baby diarrhea.

Mike: Yeah, it's kind of, it's about that color and that's one that I don't like. Yeah, it's bad, isn't it?

Nancy: That is disgusting.

你最喜欢的颜色

麦克:嘿,南西,你好啊。

南西:嘿,麦克,你好。

麦克:我想问你个问题,你最喜欢什么颜色?

南西:最喜欢的颜色?绿色吧。

麦克:哦,为什么是绿色?

南西:我不知道,这段时间我有点喜欢绿色,不过这也经常变的。

麦克:像季节变化那样?你喜欢的颜色随季节变化而变化?

南西:我觉得每个月喜欢的都不同。

麦克:好吧。

南西:那你呢,你最喜欢什么颜色?

麦克:我喜欢紫色。

南西:为什么是紫色?

麦克:我也不知道,我猜和你的原因差不多,不过我不穿紫色。我的意思是我没有紫色的衣服,我的房间也没有任何紫色的东西。

南西:你应该买一件紫色西装。

麦克:一套佐特西装,紫色的佐特西装。

南西:对的,买一套穿上,当你觉得你对紫色很有感觉的时候。

麦克:好的,我觉得这个可行。

南西:对。那你不喜欢的是什么颜色?

麦克:不喜欢的颜色?我指给你看我不喜欢什么颜色。但我不知道该怎么叫这种颜色。也许近似的颜色叫芥末色?

南西:像这样的绿色?

麦克:你觉得这叫绿?我觉得更像婴儿拉稀的屎色。

南西:你这么一说的确像屎色。

麦克:对吧,有点像吧,这就是我不喜欢的颜色,很恶心对吧。

南西:是挺恶心的。

Section B: Reading Ⅰ

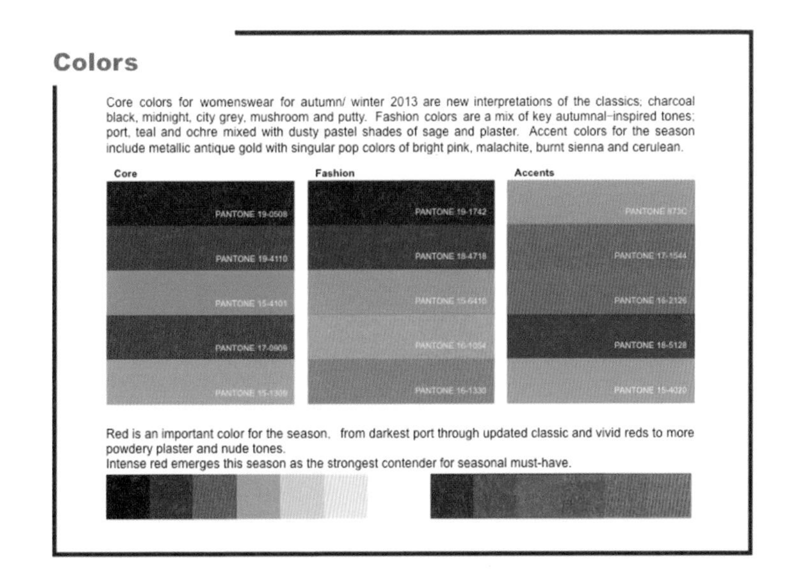

Words and expression

core　主要的

accent　强调

interpretation　理解

charcoal black　炭黑

midnight　漆黑

city grey　高级灰

mushroom　蘑菇色

putty　油灰色

autumnal inspired　秋天的灵感

port　酒红

teal　蓝绿色

ochre　赭石

dusty pastel　柔和的

sage　橘黄色

plaster　石膏色

metallic antique gold　金属的古金色

singular pop　特别流行的

bright pink　亮粉色

malachite　孔雀石

burnt sienna　熟赭

cerulean　天蓝色、蔚蓝色

Exercise Ⅰ:

According to the words and expression, translate reading Ⅰ into Chinese.
请根据已经学习的单词,将阅读Ⅰ的英文部分翻译成中文。

____

____

____

____

____

____

Section C: Reading Ⅱ

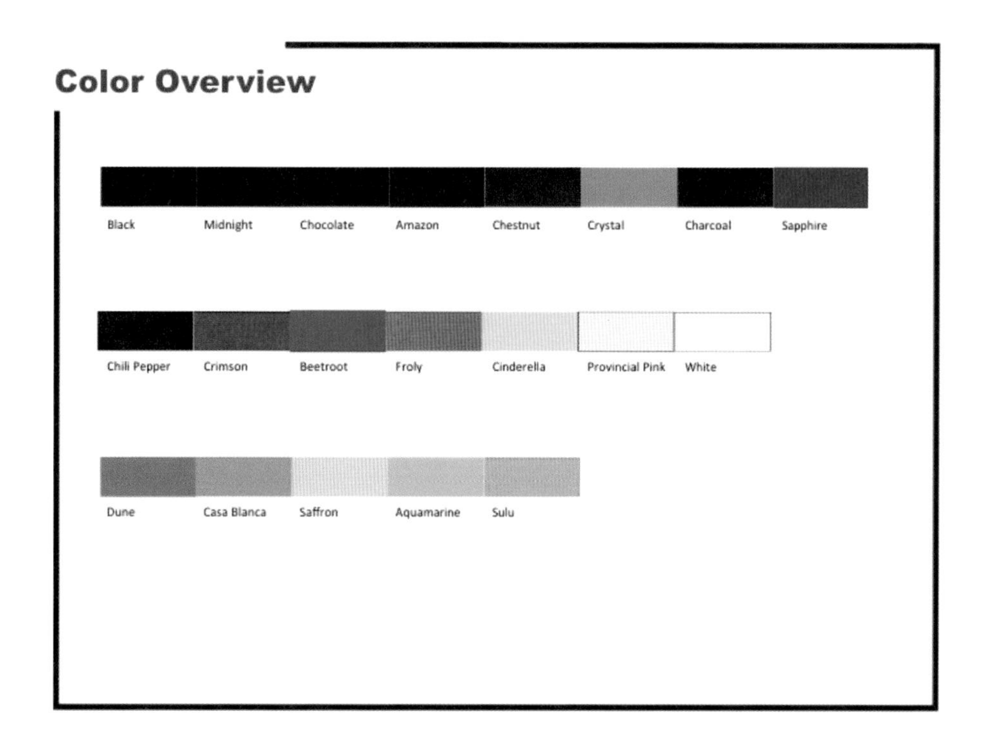

Words and expression

color overview　色彩概览

chocolate　巧克力色

amazon　丛林绿

chestnut　栗色

crystal　水晶

charcoal　木炭;深灰色

sapphire　蔚蓝色;宝石蓝

chili pepper　红辣椒色

crimson　深红色

beetroot　红甜菜根

froly　粉红的橙色

cinderella　灰粉色

provincial pink　旧粉色

white　白色

dune　沙丘

casa blanca　香水百合

saffron　番红花黄(介于嫩黄和砖黄之间)

aquamarine　海蓝色

sulu　苏禄绿

Exercise Ⅱ:

According to the words and expression, collect the color words that are not mentioned upside, and select several colors you like and write down the reasons you like them. 请根据已经学习的单词,收集上面图表中没有提到的有关色彩的单词,选择几个你喜欢的颜色单词并写下你喜欢这个颜色的原因。

____

____

____

____

____

____

Module Two —— Apparel forecasting Ⅰ 服装预测Ⅰ

Section A: Speaking

Fashion Style

Shelley: Hey, Lindy. How are you doing?

Lindy: Hey, Shelley. I am good. These pants look unique. What's the brand?

Shelley: The brand? I don't know, but there's a story behind the pants.

Lindy: Oh, yeah, what's the story behind the pants?

Shelley: Last weekend, I was wondering the shopping mall. These pants got heavily discounted, so I purchased them and then went home. Yesterday, when I took out these pants, I was like, "Wait a second! I know why no one else would buy those cause they're so ugly."

Lindy: What, you said these pants are ugly?

Shelley: Yeah. No, actually, I think my first thing is I look at the price tag and I was like, "Ah, dam, 60 RMB, I could wear as work pants." And then I bought them.

Lindy: Well, but these pants suit you very well, and they look unique!

Shelley: Really? you really think so?

Lindy: Of course. The greatest thing about being a model is you can wear whatever you want. And these pants suit you very well.

Shelley: Thank you.

Lindy: I just think people should choose the styles which suit them very well rather than the expensive ones.

Shelley: For sure.

Lindy: You should wear it.

Shelley: OK. Thank you.

服装款式

雪莉:嘿,林迪,最近怎么样?

林迪:嘿,雪莉,我挺好的。你这条裤子看起来挺奇特的,是什么牌子的?

雪莉:牌子?我不知道,但是我买这条裤子还有个故事呢。

林迪:哦,真的吗,什么故事?

雪莉:上个礼拜我在逛店,这些裤子打折力度很大,所以我就买了几条然后回家了,昨天,我把这些裤子拿出来,我反应过来:“等一下,我好像知道为什么没有人买这些裤子了,因为它们太丑了”。

林迪:什么,你觉得这些裤子很丑?

雪莉:对的。哦,不是,事实上我买裤子的时候首先看了一下吊牌,“天啊,只要60元,我当工作裤穿都值了”。所以我买了这些裤子。

林迪:好吧,但是我觉得这条裤子很适合你,而且看起来很独特。

雪莉:真的吗?你真的这样认为?

林迪:当然啦,你们模特最大的优点就是穿什么都好看,想穿什么就穿什么。而且这条裤子真的很适合你。

雪莉:谢谢。

林迪:而且我觉得人们在挑选衣服款式的时候都应该挑选适合他们的而不是贵的。

雪莉:那倒是真的(我很同意)。

林迪:你穿上挺好看的,好好穿着。

雪莉:好的,谢谢。

Section B: Reading Ⅰ

Interaction, communication and user experience inspire a mix of human, animal and technological elements. Hybrid design mirrors the new relevance of comfort and protection through quilted and wrapped constructions. Flowing organic lines and dipped coatings exhibit aerodynamic or seamless properties. Products take on the look of creatures, have friendly artificial characteristic and become personalities in their own right.

Words and expression

interaction 互动

user experience 用户体验

inspire 激励

hybrid 混合

mirror 反映

relevance 现实主义

protection 保护

quilted 夹棉的

wrapped 包裹的

construction 结构

flowing 流动的

organic 有机的

lines 线条

dipped 浸染的

exhibit 展现

aerodynamic 空气感

seamless property 无缝属性

Exercise Ⅰ:

According to the words and expression, translate reading Ⅰ into Chinese.
请根据已经学习的单词,将阅读Ⅰ的英文部分翻译成中文。

Section C: Reading Ⅱ

In a world of extremes and turbulent unrest, neutrality becomes the most progressive standpoint. Shifts in perception, extreme androgyny and a faux uniformity inform this new minimalism that is soft, immersive and boldly modest.

Words and expression

intrinsic　本质的

epicene　中性的

neutrality　中立状态

standpoint　观点

androgyny　雌雄同体

uniformity　均匀性

inform　预示

new minimalism　新的极简派

Exercise Ⅱ:

According to the words and expression, translate reading Ⅱ into Chinese.
请根据已经学习的单词,将阅读Ⅱ的英文部分翻译成中文。

Module Three —— Apparel forecasting Ⅱ 服装预测Ⅱ

Section A: Speaking

Fashion Show

Nina: Hi, Lucas, I'd like to talk about the preparation of the fashion show if you're free now.

Lucas: Hi, Nina, how about twenty minutes? I have a meeting half an hour later.

Nina: Ok. That's enough. Firstly, I wanna know, what's the topic of this fashion show?

Lucas: The topic for the S/S fashion collection this year would be Sky.

Nina: Got it. That's cool, so I guess the main color should be blue.

Lucas: What about style for the runway? Any good idea?

Nina: It can be a little bit special. For example, we can have some Seagull and some cloud models inside, to respond to our theme.

Lucas: It sounds good. Will the budget be high?

Nina: How much is the budget?

Lucas: About 150,000RMB.

Nina: That will be enough. For this kind of fashion show, expenses are mainly on the rental, stage set-up, model, advertising, music and lighting, and labor cost. How many styles to be shown?

Lucas: About 90 styles.

Nina: Then 15 models would be enough.

Lucas: Ok. You decide it.

Nina: Anything else should I notice?

Lucas: Just choose some vibrant and young models.

Nina: No problem.

Lucas: Oh, I have to go now, see you later.

Nina: See you.

时 装 秀

妮娜:你好,卢卡斯,如果你有空的话,我想和你讨论一下服装走秀的准备工作。

卢卡斯:你好,妮娜,20分钟可以吗,因为我半个小时后有个会议?

妮娜:好的,时间足够了。第一件事,我们今年时装秀的主题是什么?

卢卡斯:今年春夏时装秀的主题是天空。

妮娜:好的。是个很酷的主题,那我们的主色调就定为蓝色。

卢卡斯:今年秀场采用什么风格,你有什么建议吗?

妮娜:我觉得可以稍微特别一点。比如说,我们可以在中间放一些海鸥和云朵造型的模型,与我们的主题相呼应。

卢卡斯:听起来很有创意,那预算会很高吗?

妮娜:预算是多少?

卢卡斯:大概15万人民币。

妮娜:那应该够了。这种类型的服装秀的花费主要集中在租场地的租金,舞台搭建,模特费用,广告费用,音乐和灯光费用和劳动力成本。我们有几套服装?

卢卡斯:大约90套。

妮娜:那15个模特够了。

卢卡斯:好的,你来决定。

妮娜:还有什么我需要特别注意的吗?

卢卡斯:挑模特的时候尽量挑选有活力的年轻模特。

妮娜:没问题。

卢卡斯:哦,我必须走了,待会见。

妮娜:再见。

Section B: Reading Ⅰ

Hack-tivate elevates the traditional ideas of DIY, deconstruction, customisation and repair. Unexpected styling and layering create a sense of eccentricity, while inventive cutting and color-blocking give technical elements a considered look. Quirky futuristic finishing lends cyber edge, while mending can add a personalised feel.

Words and expression

hack-tivate　黑客入侵

elevate　提升

deconstruction　解构

customisation　量身定制

repair　修理

unexpected styling　意想不到的款式

inventive cutting　独出心裁的裁剪

color blocking　色块

quirky　古怪的、诡异的

futuristic　未来感

cyber edge　网络感

mending 缝补

Exercise Ⅰ:

According to the words and expression, translate reading Ⅰ into Chinese.
请根据已经学习的单词,将阅读Ⅰ的英文部分翻译成中文。

Section C: Reading Ⅱ

Luxury takes on a brutal and sometimes delicate harmony, as we explore the dark, subterranean side of nature. Foraged food, undergrowth, culled fur and genetically engineered plant hybrids combine with a neo-shamanic perspective and a mystical aura.

Words and expression

luxury 奢侈

take on 呈现

brutal 冷酷

delicate harmony 微妙的和谐

subterranean 隐藏的

foraged food 饲料

undergrowth 矮树丛

culled fur 被剔除的毛

genetically engineered plant 转基因植物

hybrid 杂种的、混合的

neo-shamanic perspective 新的萨满教观点

mystical aura 神秘的光环

Exercise Ⅱ:

According to the words and expression, translate reading Ⅱ into Chinese.
请根据已经学习的单词,将阅读Ⅱ的英文部分翻译成中文。

Section D: Reading Ⅲ

The Classically Romantic Themes, which surround craftsmanship, luxury, and the pursuit of beauty, are updated with a contemporary sharpness, delicate sheers and trims combine with dance wear in a modern take on fantasy, humble draping and simple silhouettes feel clerical, while intricate details, added treatments, and ornate baroque costumes add a theatrical sense of ceremonial grandeur.

Words and expression

craftsmanship 精湛技艺

luxury 奢华

contemporary sharpness 当代的犀利

update 更新

delicate sheers 精致的薄纱

dance wear 舞蹈服装

humble draping 柔和的垂坠

clerical 牧师的

intricate 复杂的

added treatment 做了很好的补充

ornate baroque 华丽的巴洛克

theatrical 戏剧的

ceremonial grandeur 仪式的庄严

Exercise Ⅲ:

According to the words and expression, translate reading Ⅲ into Chinese.
请根据已经学习的单词,将阅读Ⅲ的英文部分翻译成中文。

Module Four —— Graphic & print 图案和印花

Section A: Speaking

Fashion designer

Jenny: Nice to meet you! I'm Jenny, a journalist of ELLE. We made an appointment yesterday. Can I have some quick question for you?

Alex: Ah, Nice to meet you too. Yes, of course.

Jenny: Ok. Shall we start now?

Alex: Yeah.

Jenny: Ok. Let's come directly to our topic. In your opinion, what is clothing?

Alex: Personally, I think it is my youth and my enthusiasm.

Jenny: Oh, It's so cool. And how about fashion?

Alex: Eternity, even if not, but still look forward to it.

Jenny: So how about design?

Alex: My exclusive dream that I will never give up.

Jenny: So cool. Next one. How do you get along with your companions in your department?

Alex: Besides the designer, you will find pattern cutter and sample machinist here. To some degree, a designer is a leader to head a small team, closely cooperating with each other.

Jenny: OK, then, how do you define a successful designer?

Alex: It sounds different from different people. For me, I think a designer must be creative, independent, with smart insights into fashion, as well a strong sense of market.

Jenny: Have you done these?

Alex: I'm very sorry. I'm sure that I'm a good designer but not successful.

Jenny: Oh. You're too modest. That's all for my questions. Thanks a lot!

Alex: It's my pleasure.

服装设计师

杰妮:很高兴见到你,我是杰妮,ELLE 的记者。我们昨天约好今天见面的。我们可以有一个快速采访吗?

艾利克斯:啊,很高兴见到你。是的,当然可以。

杰妮:那我们现在开始好吗?

艾利克斯:好的。

杰妮:那我们直接进入主题吧,在你看来,服装是什么?

艾利克斯:就我个人而言,我觉得服装是我的青春和我的热情。

杰妮:哦,很有趣。那你觉得时尚是什么?

艾利克斯:时尚是永恒,即使不能,但我也仍然盼望如此。

杰妮:那你对设计是怎么看的?

艾利克斯:是专属于我的,永不放弃的梦想。

杰妮:太酷了,下一个问题,在设计部门,你如何与同事相处?

艾利克斯:在设计部门,除了设计师,还有制版师和样衣工。某种程度上说,设计师就是一个小组的领导,和每一个成员密切合作。

杰妮:好的,那你怎么定义一名成功的设计师呢?

艾利克斯:每个人有每个人的看法,就我而言,我觉得一个成功的设计师必须非常有创造力,独立,对时尚有敏锐的洞察力,还要有很强的市场感。

杰妮:那你做到了吗?

艾利克斯:不好意思,我能确定自己是一个好设计师,但并不是成功的设计师。

杰妮：哦。你太谦虚了。我的问题问完了，谢谢你。
艾利克斯：不客气。

Section B: Reading Ⅰ

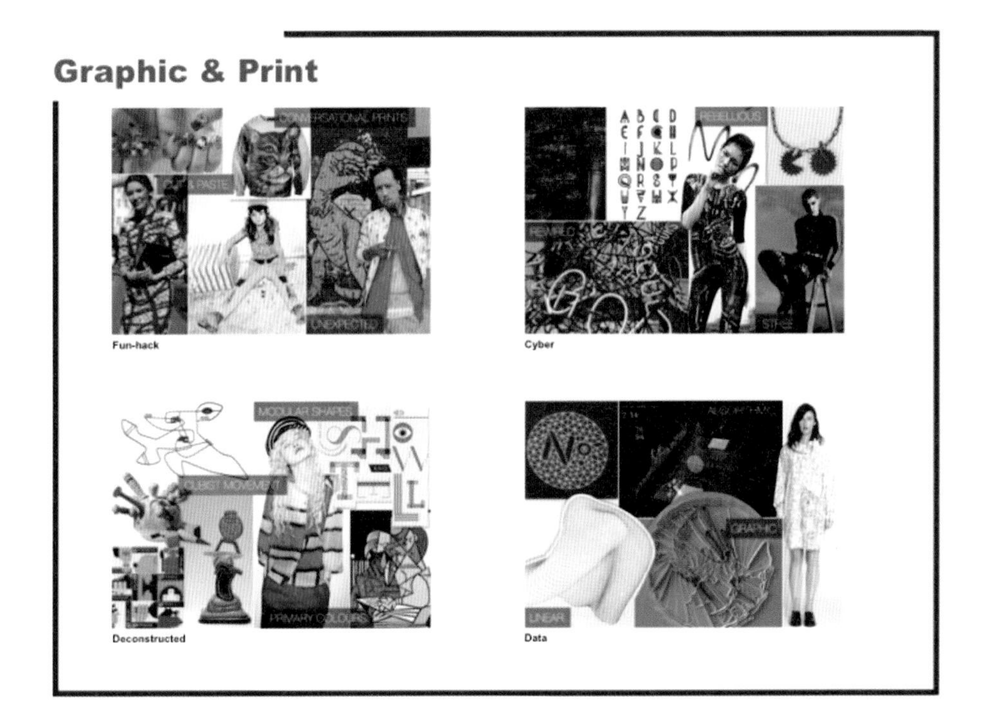

Words and expression

graphic & print 图案和印花
fun-hack 有趣的黑客
cyber 网络
deconstructed 解构
data 数据

Section C: Reading Ⅱ

Words and expression

ethereal　空灵

dark floral　黑色的花

new classical　新经典

chameleon　变色龙

Exercise Ⅰ:

According to the words and expression, describe the graphic & print you like in English. 请根据已经学习的单词,用英语描述你喜欢的图案或印花风格。

Section D: Reading Ⅲ

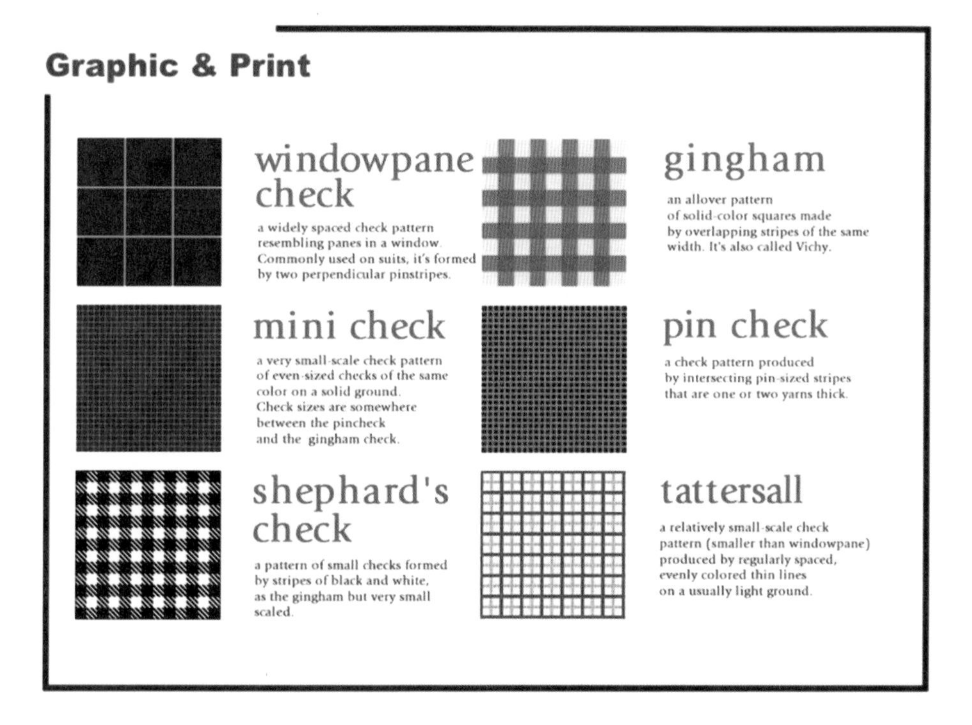

Words and expression

windowpane checks 窗格纹

gingham 方格色织布

mini check　迷你方格纹

pin check　细格子纹

shephard's check　黑白小格子花纹

tattersall　塔特萨尔花格(浅底深色方格)

Section E: Reading Ⅳ

Graphic & Print

gun club check

it was originally known as "The Coigach" and developed in Scotland. District checks were used as a uniform of sorts for Scottish estates.

houndstooth

Aka pied-de-poule, it is a duotone textile pattern characterized by broken checks or abstract four-pointed shapes, often in black and white.

harlequin check

a pattern with diamond shaped checks. It can come in many colors and dimensions.

graph check

a check pattern created by crossing lines on a solid ground that resembles graph paper.

glen check

also known as the Prince of Wales, it's the district check of the Glen Urquhart esate and combines different sizes of checks and houndstooth.

argyle

this pattern is made of diamonds or lozenges. Typically, there is an overlay of intercrossing diagonal lines

Words and expression

gun club check　GC格纹(有好多颜色)

houndstooth　千鸟格

harlequin check　菱形花纹格子

graph check　十字图纹格

glen check　格伦格子

argyle　阿盖尔菱形花纹

Exercise Ⅱ:

Collect the graphics and prints of 3 famous brands, write down your feelings of the graphics and prints and how you will use them in your own design. 收集3个知名品牌的图案和印花,写下这3款图案和印花给你的感觉和你将如何运用这些图案在自己的设计中。

____

____

____

____

____

____

Project Two Fashion Design and Styling
服装设计与造型配饰

Module One —— Silhouette & Details Ⅰ
廓形和细节Ⅰ

Section A: Speaking

Fashion style Ⅰ

Cindy: Hi, Grace! Your dress is so beautiful.

Grace: Oh, Cindy. I'm flattered.

Cindy: The long skirt makes you appear much longer and thinner. It's really looks good on you.

Grace: Thank you. I like the open bottom of the skirt especially.

Cindy: Yes. The silhouette is really classic. By the way, what's the brand?

Grace: The brand is ABC.

Cindy: Oh, I know the brand, and the dress suits you very much.

Grace: Thank you for your praise. You look nice too! When did you change your wearing style and dress the business uniform?

Cindy: You notice that! Do you think I fit for the beige color uniform?

Grace: You are a pretty girl and suit for any clothes.

Cindy: I think there will be more fresh feeling when you changing your wearing style sometimes.

Grace: Absolutely!

Cindy: By the way, Do you have time to go shopping with me this weekend?

Grace: Oh, it's a great idea. I'm free this weekend.

Cindy: Ok, See you then.

Grace: See you.

服装款式Ⅰ

辛迪:你好,格丽斯。你的裙子好漂亮啊。

格丽斯:哦,辛迪。我可要飘飘然了。

辛迪:长裙显得你又高又瘦。你穿真的好看。

格丽斯:谢谢,我特别喜欢裙子下摆的开口。

辛迪:是的,裙子的廓形真的很经典。对了,这是什么牌子?

格丽斯:裙子的品牌是 ABC 的。

辛迪:哦,我知道这个牌子,这裙子实在太适合你了。

格丽斯:谢谢你的夸奖,你看起来也很不错啊,你什么时候改变穿衣风格开始穿职业套装了?

辛迪:你注意到了!你觉得我适合这种米色套装吗?

格丽斯:你是美女,适合任何衣服。

辛迪:我觉得有时候换一下穿衣风格能带来许多新鲜感。

格丽斯:确实是这样。

辛迪:对了,这个周末你有空和我一起逛街吗?

格丽斯:哦,这真是个好主意。我这周末有空。

辛迪:好的,那到时候见啦。

格丽斯:再见。

Section B: Reading Ⅰ

Silhouette and Details

Coat/ Trench: retro cocoon; tulip; oversized Crombie; minimal trench

Jacket/ Blazer: slim boyfriend; schoolboy blazer; sleeveless tux; funnel-neck puffs; sports blouson; minimal biker; technical hybrid

Dress: fit & flare; bubble hem; woven body-con

Top: vintage blouse; boyish shirt; couture tee; raglan sweatshirt

Jersey: slim-fit polo; asymmetric drape jersey; drop-back sweatshirt; simple jersey shirt

Skirt: mini bubble; gentle flute

Pants: tailored volume; flat-front straight leg; woven trackie; low-slung crop

Sweater: chunky cardigan coat; boxy retro cardigan; cascade drape cardigan; skinny rib knit; cowl-neck tabard

Words and expression

trench　风雨衣

jersey　运动服

jacket　夹克

skirt　短裙

pants　裤子

dress　连衣裙

sweater　毛衣

top　上衣

Section C: Reading Ⅱ

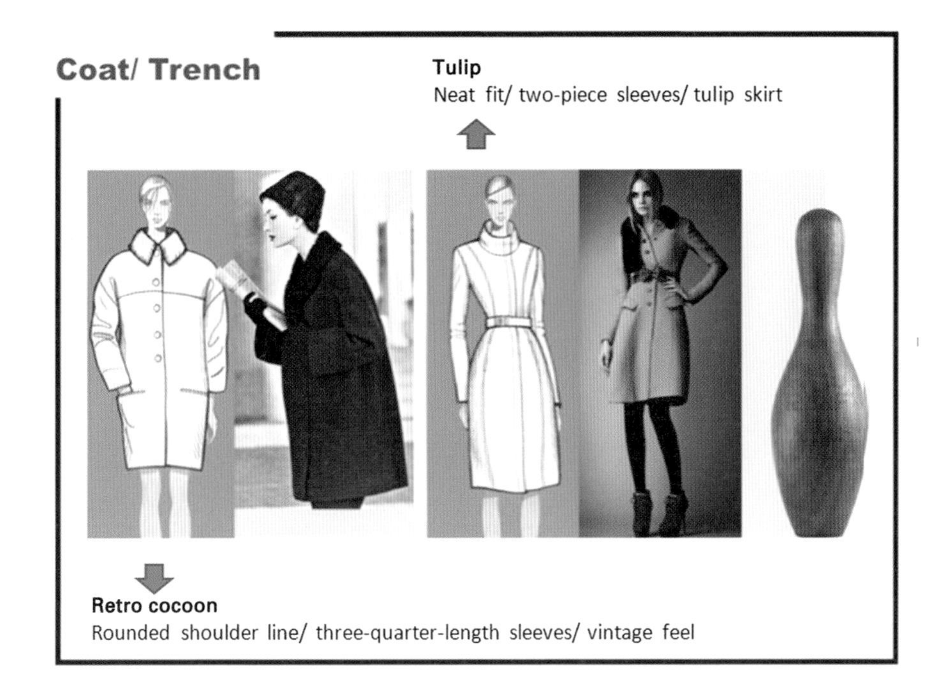

Words and expression

retro cocoon 复古茧型

rounded shoulder line 圆肩线

three-quarter-length sleeves 七分袖

vintage 复古

tulip 郁金香型

neat 整洁、利落

two-piece sleeve 两片袖

Section D: Reading Ⅲ

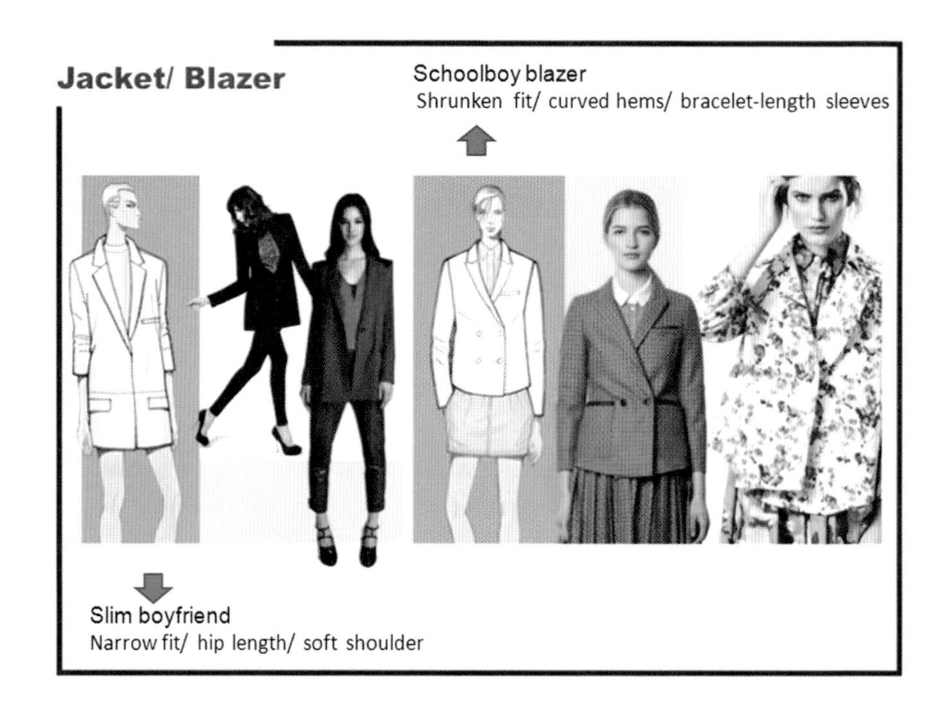

Words and expression

slim boyfriend 修身男朋友外套

narrow fit 窄版

hip length 臀部长度

soft shoulder 柔软肩部

schoolboy blazer 学院风西装

shrunken fit 小版

curved hems 弧形下摆

bracelet-length 手腕长度

Section E: Reading Ⅳ

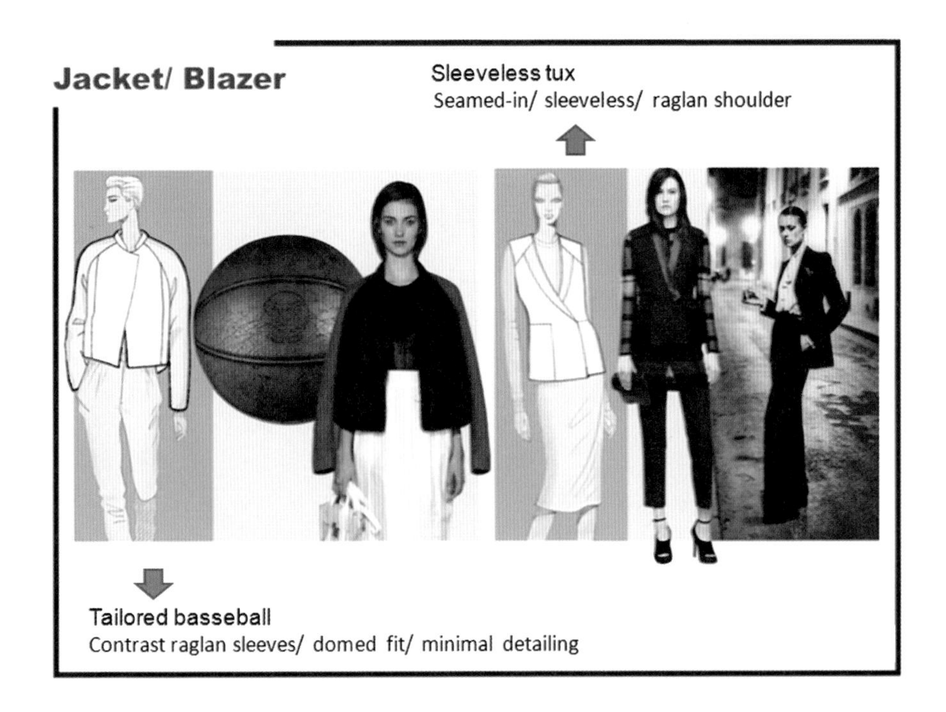

Words and expression

tailored baseball 精致球衫

contrast raglan sleeves 对比插肩袖

domed fit 半球形廓型

minimal 最小的，最少的

sleeveless tux 西装背心

seamed-in 内部有缝线

raglan 插肩的

Section F: Reading Ⅴ

Words and expression

fit & flare　合身的喇叭裙

full skirt　长裙

fitted bodice　合体的

bubble hem　泡泡臀

dropped waist　低腰的

skinny　紧身的

shift update　变换更新

Section G: Reading Ⅵ

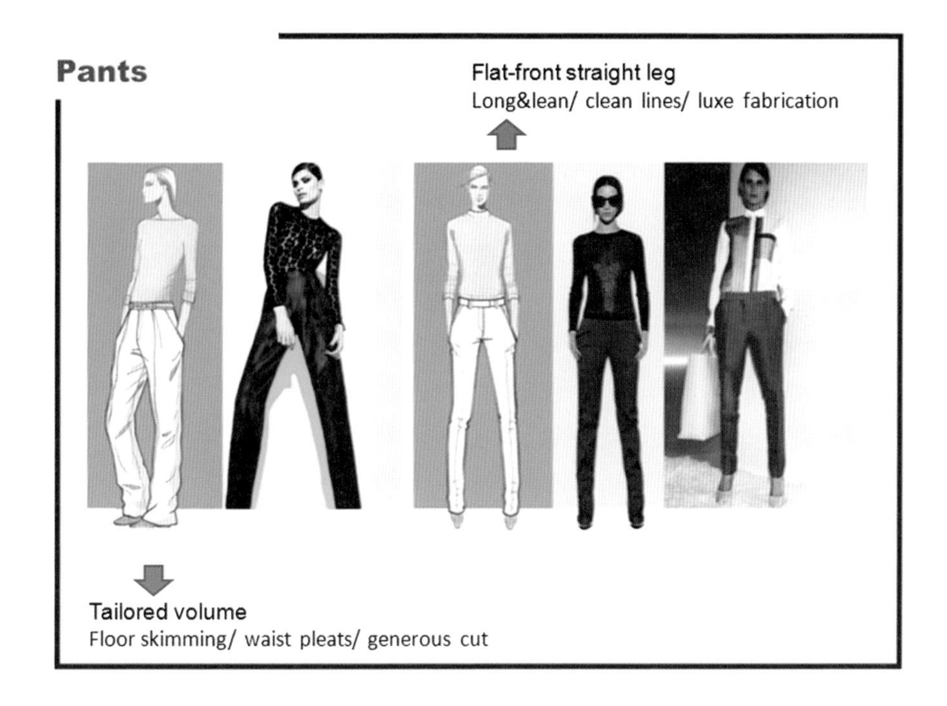

Words and expression

tailored volume　精致的大体积

floor skimming　长度触碰到地面

waist pleats　腰部褶裥

generous cut　剪裁大方

flat-front straight leg　前平落裆直腿裤

long & lean　细长的

clean lines　简洁的线条

luxe fabrication　奢华面料

Section H: Reading Ⅶ

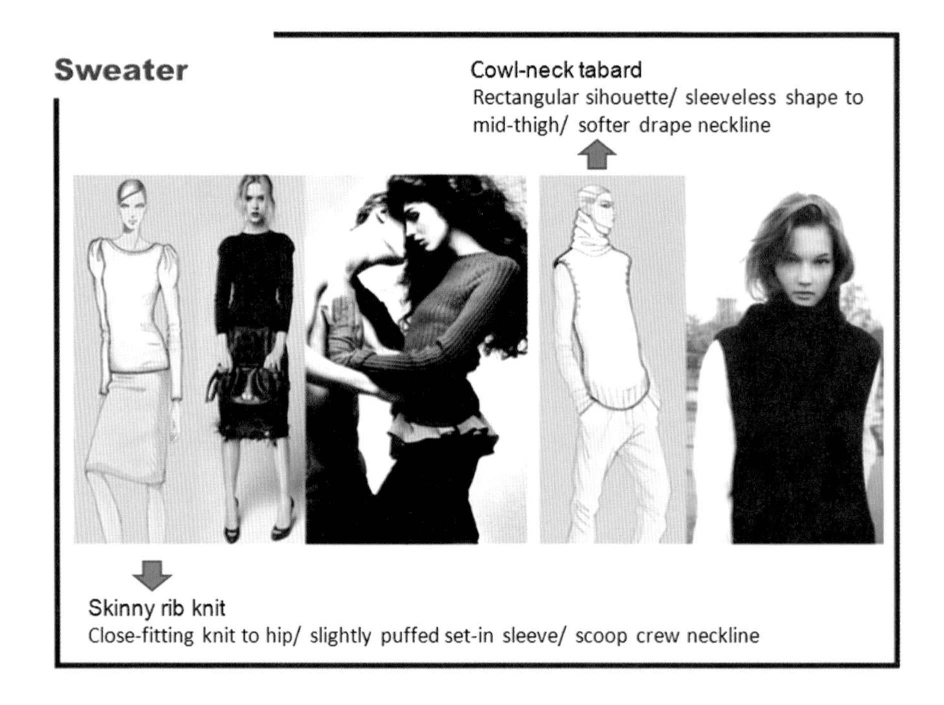

Words and expression

skinny rib knit　紧身针织罗纹

close-fitting knit to hip　紧身中长针织衫

slightly　略微、稍微

scoop crew neckline　勺型圆领(低圆领)

cowl-neck　高翻领

tabard　高领背心(无袖短外套)

rectangular　矩形

mid-thigh　大腿中部

softer　柔软、舒适

drape neckline　垂坠领

Exercise:

According to the words and expression, describe the clothes you wear today in English. 请根据已经学习的单词,用英语描述你今天穿着的服装。

Module Two —— Silhouette & Details Ⅱ 廓形和细节Ⅱ

Section A: Speaking

Fashion style Ⅱ

Norah: Hey. What're you looking for today?

Lisa: I need a coat but I want the coat to be very warm also very formal, so that I can wear it for the interview.

Norah: Ok. Do you have any other requirements?

Lisa: No.

Norah: Ok. Well, we have this coat here which is very beautiful and formal. Enough to wear in an interview.

Lisa: Oh. I like it. Let me try it. What size do you have?

Norah: We have size S, M and L.

Lisa: I will try the medium one.

Norah: Ok. How do you like it?

Lisa: Wow. I actually like it a lot. What do you think?

Norah: It looks great.

Lisa: Is it formal enough for the interview?

Norah: Yes. You can definitely wear it to an interview.

Lisa: Ok. How much is it?

Norah: It's RMB499.

Lisa: Wow. That's a lot. That's definitely over my budget.

Norah: Well. We actually have a promotion now. If you buy this coat, you can also get an our best-selling design for 50% off. And this is also formal enough to wear to an interview as well.

Lisa: You're such a good seller. I really wanna buy it.

Norah: Ok. Great! Two Ms.

Lisa: Yes. Definitely.

服装款式Ⅱ

诺拉:你好,你今天想买什么?

丽萨:我想买件又保暖又比较正式的外套。因为我想穿它去面试。

诺拉:好的,还有其他需求吗?

丽萨:没有了。

诺拉:好的。这件外套很漂亮又很正式,很适合面试的时候穿着。

丽萨:哦,我很喜欢,我能试一下吗?有哪些码数?

诺拉:这款有S码、M码和L码。

丽萨:给我试一下M码。

诺拉:好的。你觉得怎么样?

丽萨:哦,我很喜欢,你觉得呢?

诺拉:看起来很不错。

丽萨:面试的时候穿够正式吗?

诺拉:够了,你穿这件去面试完全没有问题。

丽萨:好的,这件多少钱?

诺拉:499人民币。

丽萨:哦,有点贵,超出了我的预算。

诺拉:我们现在有个促销活动,你买这件衣服的话,可以五折购买一件我们的畅销品,也是非常适合面试穿着的比较正式的款。

丽萨:你真是一个好销售啊,那我买了。

诺拉:好的,两件M码。

丽萨:对,没错。

Section B: Reading Ⅰ

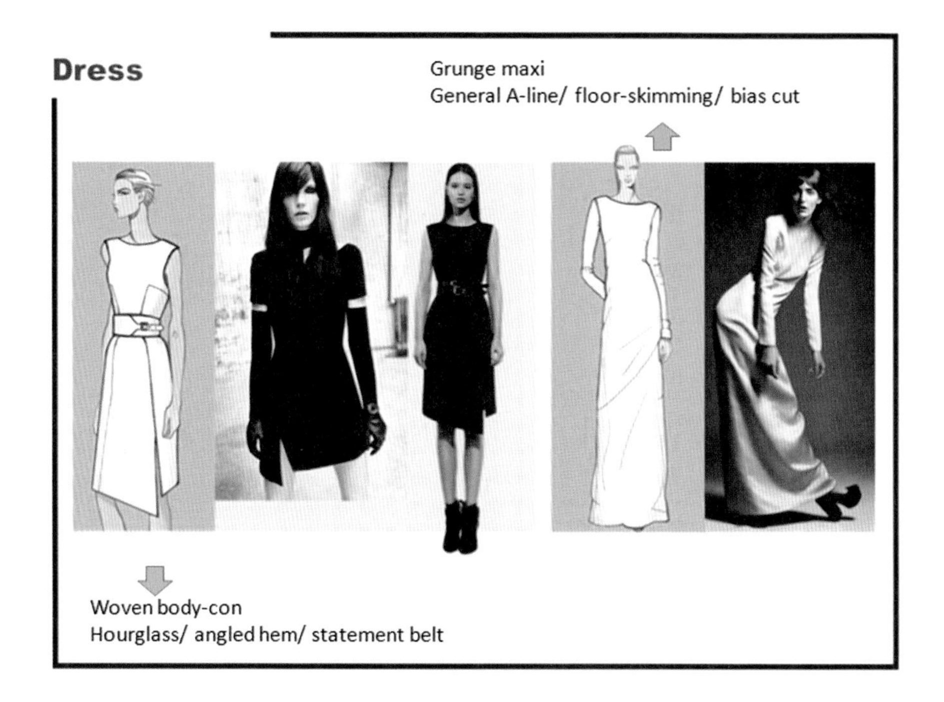

Words and expression

woven 编织

body-con 紧身连衣裙

hourglass 沙漏型

angled hem 有角度的下摆

statement belt 夸张的大腰带

grunge 摇滚风

maxi 长至脚踝的连衣裙

general a-line 大致成 a 型

floor-skimming 掠过地面的(裙长至地)

bias cut 斜裁

Section C: Reading Ⅱ

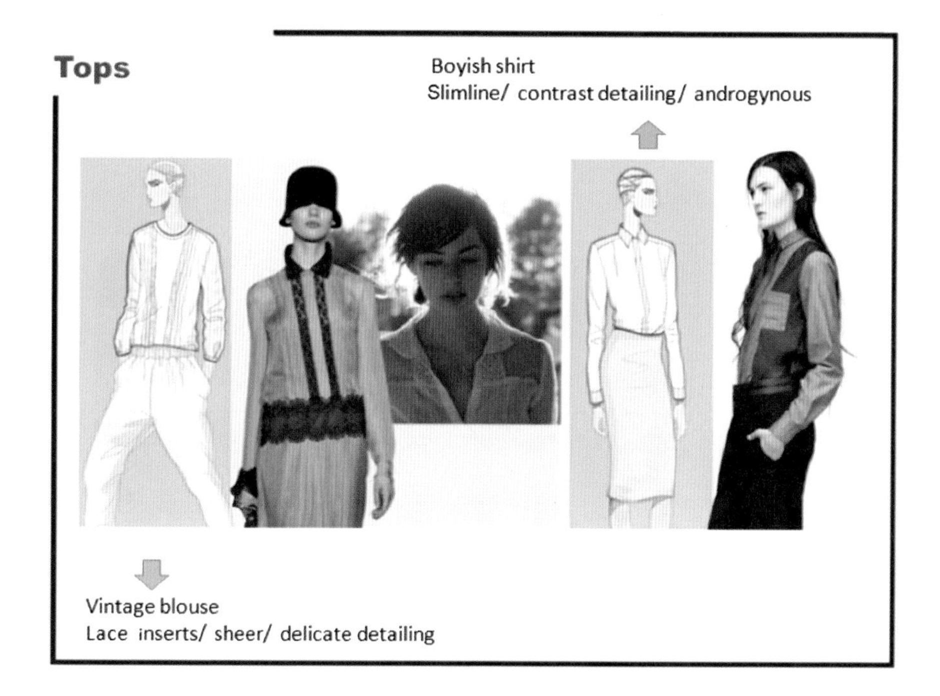

Words and expression

vintage blouse 复古女衬衫

lace inserts 嵌入蕾丝的

sheer 薄绸

delicate 精致的

detailing 细部设计

boyish shirt 男孩式的衬衫

slimline 纤细的

contrast 形成对比的

androgynous 中性的

Section D: Reading Ⅲ

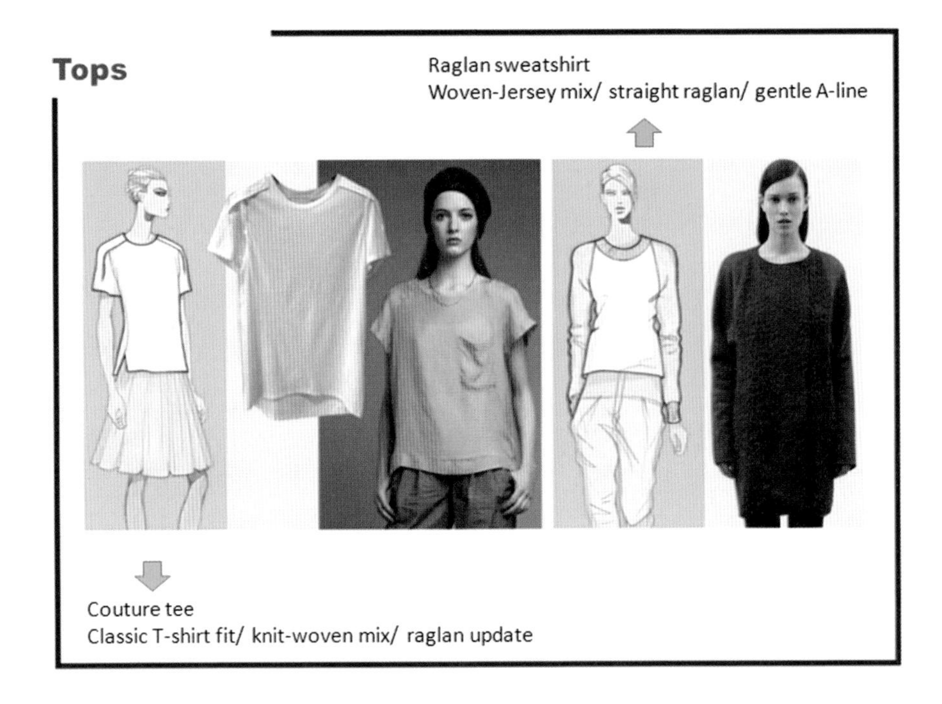

Words and expression

couture tee　高定 t 恤

classic t-shirt fit　经典 t 恤型

knit-woven mix　针织梭织混合设计

raglan update　插肩袖设计

raglan sweatshirt　插肩袖毛衫

woven-jersey mix　梭织针织混合设计

straight raglan　直线型插肩袖

Section E: Reading Ⅳ

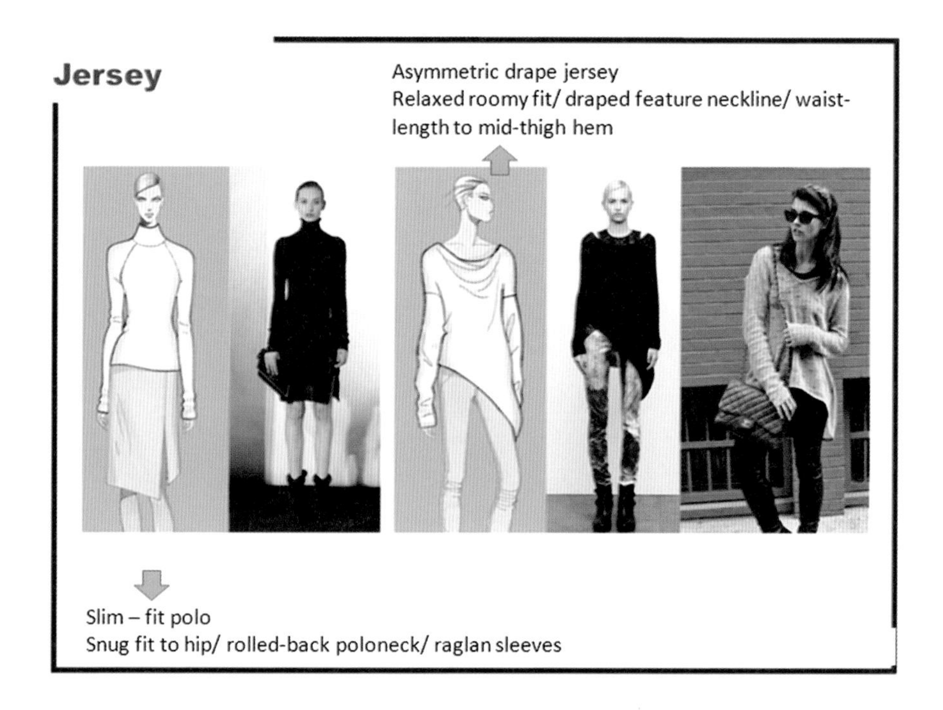

Words and expression

slim-fit polo　紧身高圆领针织衫

snug　贴身的、紧身的

rolled-back poloneck　高圆翻领

asymmetric drape jersey　不对称悬垂运动衫

relaxed roomy fit　休闲宽松型

draped feature neckline　垂荡领

waist-length to mid-thigh hem　斜下摆(从腰部到臀部)

Section F: Reading Ⅴ

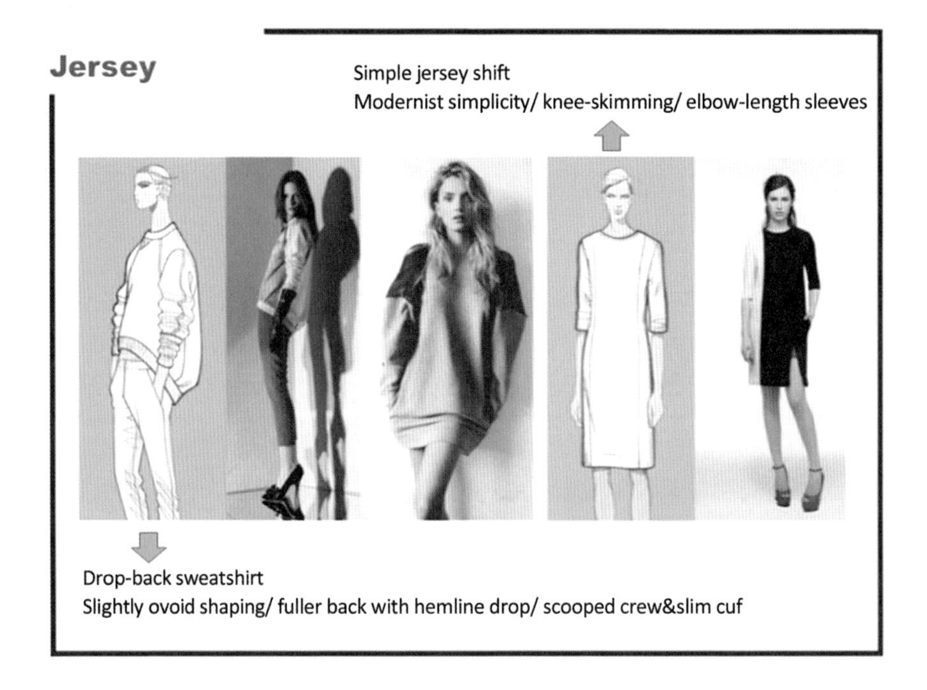

Words and expression

drop-back sweatshirt　休闲卫衣

ovoid　卵型的

fuller back with hemline drop　背部充盈下摆下垂

slim cuf　收紧克夫

simple jersey shirt　简洁针织衫

modernist　现代的

simplicity　简单

knee-skimming　长度至膝盖

elbow-length sleeves　袖长至肘部

Section G: Reading Ⅵ

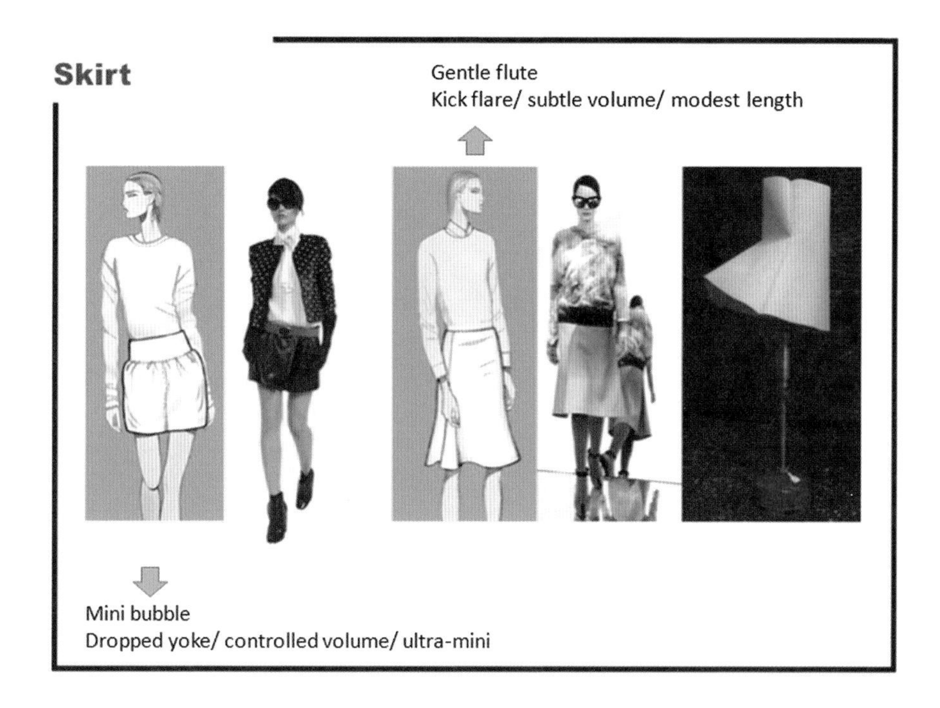

Words and expression

mini bubble 迷你泡泡裤(裙)

dropped yoke 下降育克线

controlled 有控制的

volume 体积

ultra-mini 超迷你

gentle flute 温柔的荷叶边裙

kick flare 微喇型

subtle 不明显的

modest length 中等长度

Exercise:

According to the words and expression, match 3 sets of clothes you like in English. 请根据已经学习的单词,用英语搭配3套你喜欢的服装。

____

____

____

____

____

____

Module Three —— Draping 立体裁剪

Section A: Speaking

Draping

Lucy: Hi, Lily. What are you doing now? The class is over.

Lily: Lucy, I forget the preparation step of draping that our teacher taught last week. Can you help to review it?

Lucy: Ok. My pleasure. let's begin. I remember the first step is making reference lines on a dress form.

Lily: Yes, you're right. Making reference lines. Is there anything we have to pay attention to?

Lucy: Yes, we have. There are lots of reference lines in one dress form, so we have to start from the vertical lines.

Lily: How many vertical lines?

Lucy: Vertical lines include center front line, front princess line, side seam, center back line, back princess line.

Lily: Got it. What's next?

Lucy: Then, we have to make horizontal lines. Horizontal lines are composed of neckline, bust line, waistline and hipline.

Lily: Ok. Anything else?

Lucy: Bias lines consist of shoulder seam and armhole outline.

Lily: Right, I remember, after this we can start to fix muslin or fabric to the dress form while draping.

Lucy: No, there is an important step missing, we have to confirm grain. In draping confirming whether a fabric is crosswise grain line or lengthwise grain line is very important.

Lily: Yeah, it hit me to my mind. I have to prepare for draping now. Thank you so much.

Lucy: You're welcome. See you later.

Lily: See you.

立体裁剪

露西:嗨,莉莉,你在干什么呢?课已经结束了。

莉莉:哦,露西,我忘记了上周老师讲的立裁准备步骤了,你能帮我复习一下吗?

露西:好的,我很乐意,那我们开始吧。我记得第一个步骤是在人台上贴好辅助线。

莉莉:恩,你是对的。先贴辅助线。那我需要注意一些什么吗?

露西:是的。有一些注意事项。人台上有许多辅助线,我们从纵向辅助线开始粘贴。

莉莉:纵向辅助线有几条?

露西:纵向辅助线包括前中线,前公主线,侧缝线,后中线,后公主线。

莉莉:明白了,那下一步呢?

露西:然后我们要粘贴横向辅助线。横向辅助线包括领围线、胸围线、腰围线还有臀围线。

莉莉:好的,还有么?

露西:斜向辅助线包括肩线、袖窿圈线。

莉莉:好的。我记得粘贴辅助线之后我们就要开始固定立裁需要用到的面料或白胚布了。

露西:不对,你还忘记了一个重要的步骤。我们必须确定丝缕线。立裁的时候确定面料是直丝缕还是横丝缕是非常重要的。

莉莉:对的。我想起来了。我现在要去准备立裁了。太谢谢你了。

露西:不客气。回头见。

莉莉：再见。

Section B: Reading Ⅰ

Making reference lines on a dress from reference lines must be made in black or red adhesive band. The reference lines stuck may be vertical, horizontal and bias. Vertical lines include center front line, front princess line, side seam center back line, back princess line. Horizontal lines are composed of neckline, bustline, waistline, and hipline. Bias lines consist of shoulder seam and armhole outline.

Words and expression

reference line　参考线

adhesive band　粘胶带

stuck　粘住

vertical　垂直的

horizontal　水平的

bias　斜丝缕

center front line　前中线

front princess line　前公主线

side seam　侧缝

center back line　后中线

back princess line　后公主线
neckline　领围线
bust line　胸围线
waistline　腰围线
hipline　臀围线
shoulder seam　肩线
armhole outline　袖窿外线

Exercise Ⅰ:

According to the words and expression, translate reading Ⅰ into Chinese.
请根据已经学习的单词,将阅读Ⅰ的英文部分翻译成中文。

Section C: Reading Ⅱ

The keep of the dress from. The dress form to which reference lines have been clung should be packed in muslin so that it and its reference lines can be well protected. Using pins to fix cloth being adept at using pins as contemporary stiching is one of the basic skills in draping.

Words and expression

dress form 人台

clung 附着于(cling 的过去式)

pack 包装、包裹

muslin 白坯布

protected 被保护

pins 大头针

fix 固定

cloth 布料

be adept at doing 擅长于

contemporary 现代的

stitching 缝合、针脚

Exercise Ⅱ:

According to the words and expression, translate reading Ⅱ into Chinese.
请根据已经学习的单词,将阅读Ⅱ的英文部分翻译成中文。

Module Four —— Fabric 面料

Section A: Speaking

Fabric and Cost

June: Miss White, this is the first time for me to purchase some fabrics. May I get some information in this field?

White: Sure. First, you need to know different kinds of fabric, and try to distinguish them and use them for various effects.

June: What's kind of fabric are suitable for our collection?

White: For the pant sets, denim or twill is a good choice.

June: Sounds good.

White: By the way, are all the accessories for this collection ready?

June: Half of them haven't arrived yet. I feel worried.

White: What's wrong?

June: The logistics company delayed because of the weather, but they replied that they would deliver the goods this week as soon as possible.

White: I hope the weather will be better. And the cost has increased a lot, right?

June: Right. The boss specially mentioned about the cost control for the fabric and accessories. We need to see in which part we can lower the cost.

White: Not for the fabric. Now price for the raw material increased sharply. Or we can think ways out from lining and interlining. I will discuss with the boss again next time about this.

June: I agree with you. It will not work if to lower the quality of our product. Let's keep looking for the fabric.

White: Ok, let's go.

面料和成本

朱:怀特小姐,这是我第一次去采购面料,我能在这个领域再多了解一些信息吗?

怀特:当然。首先你需要了解各种面料,然后能区分它们,并且把它们做出各种各样的效果。

朱:哪种面料比较适合我们这次的系列?

怀特:对于裤子系列,牛仔或者全棉斜纹面料是比较好的选择。

朱:听起来不错。

怀特:对了,我们这次系列里的饰品都已经到货了吗?

朱:还有一半没到,我也有点担心这个。

怀特:什么情况?

朱:物流公司因为天气原因延误了货物派送,但他们回复说这周会尽快派送货物。

怀特:希望天气尽快好起来。据说现在成本增加了很多,是这样吗?

朱:是的。上次老板特意提到面料和饰品的成本控制问题。我们需要考虑一下我们在哪些方面可以降低成本。

怀特:不要在面料方面。现在原材料的价格增长了很多。也许我们能在里料和衬料方面想想办法。我下次有机会还会和老板谈谈这个事。

朱:我同意你的想法。降低我们产品的品质不是办法。我们还是继续找面料吧。

怀特:好的。走吧。

Section B: Reading Ⅰ

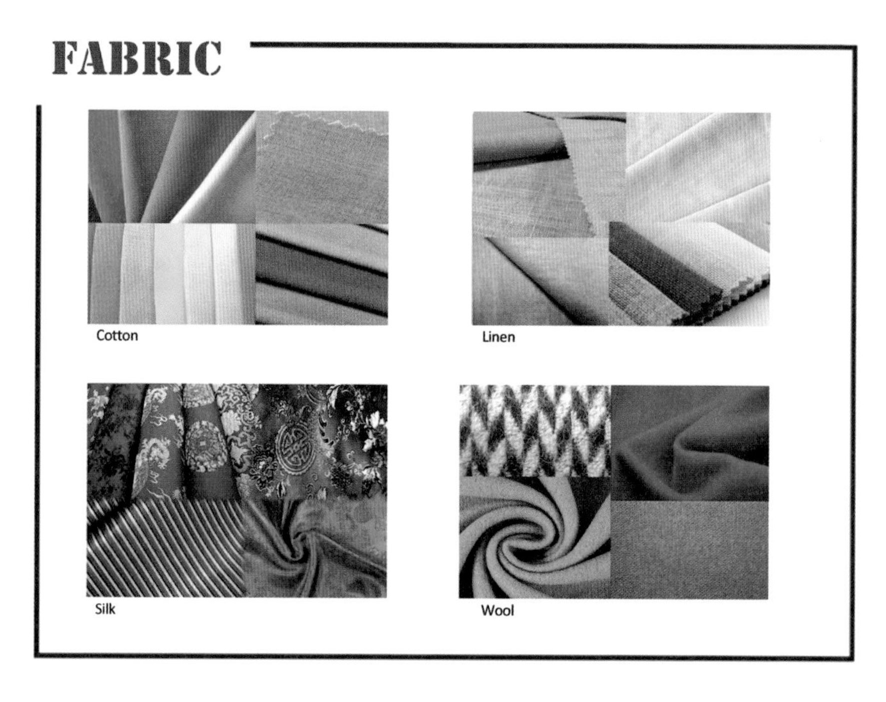

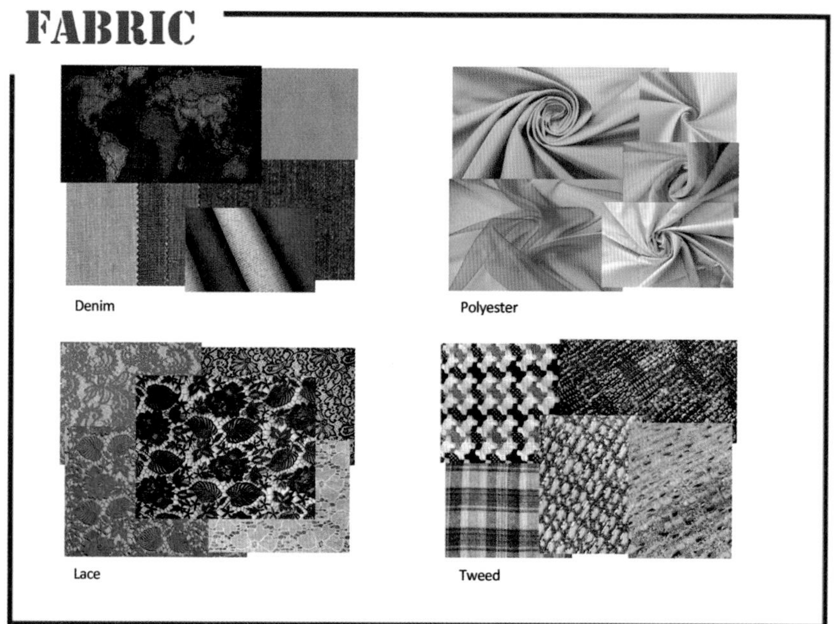

Words and expression

cotton　棉

linen　麻

silk　丝

wool　毛

denim　牛仔布

polyester　涤纶

lace　蕾丝面料

tweed　粗花呢

Exercise Ⅰ:

According to the words and expression, collect the fabric words that are not mentioned upside, select 3 kinds of fabrics you like most in your daily design, and write down the reasons why you like to use this kind of fabric. 请根据已经学习的单词,收集上面图表中没有提到的有关面料的单词,选择 3 种你平时设计中最喜欢用到的面料,并写下你喜欢用这种面料的原因。

Section C: Reading Ⅱ

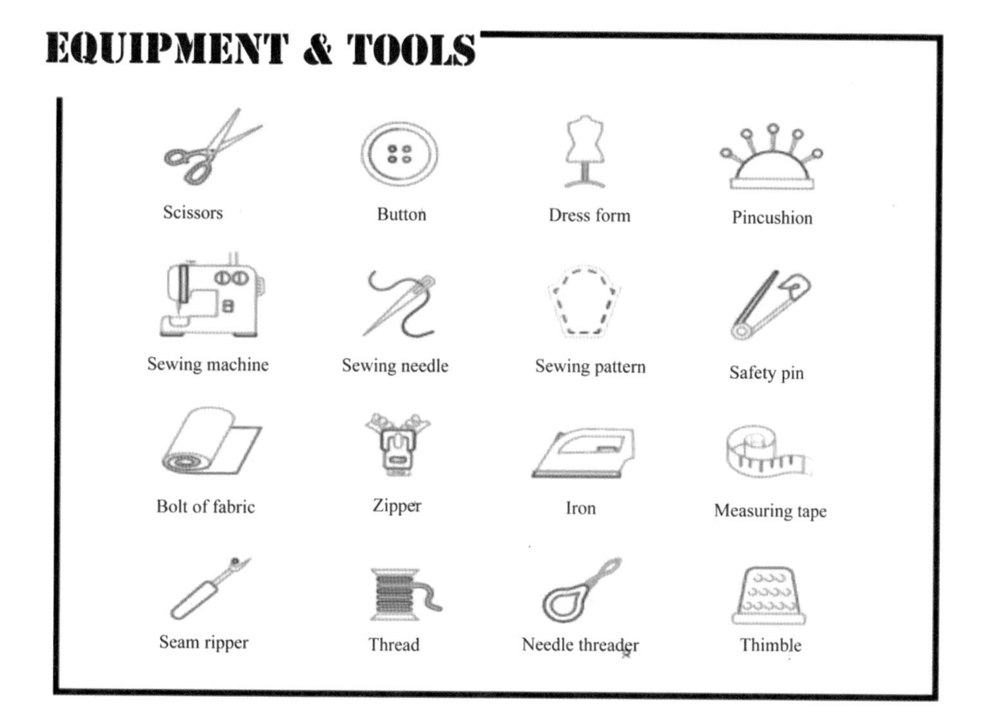

Words and expression

scissors 剪刀

button 纽扣

dress form 人台

pincushion 针插

sewing machine 缝纫机

sewing needle 缝纫针

sewing pattern 工艺样板

safety pin 别针

bolt of fabric 一卷布

zipper 拉链

iron 熨斗

measuring tape 卷尺

seam ripper 拆线器

thread 线圈

needle threader 穿线器

thimble 顶针

Section D: Reading Ⅲ

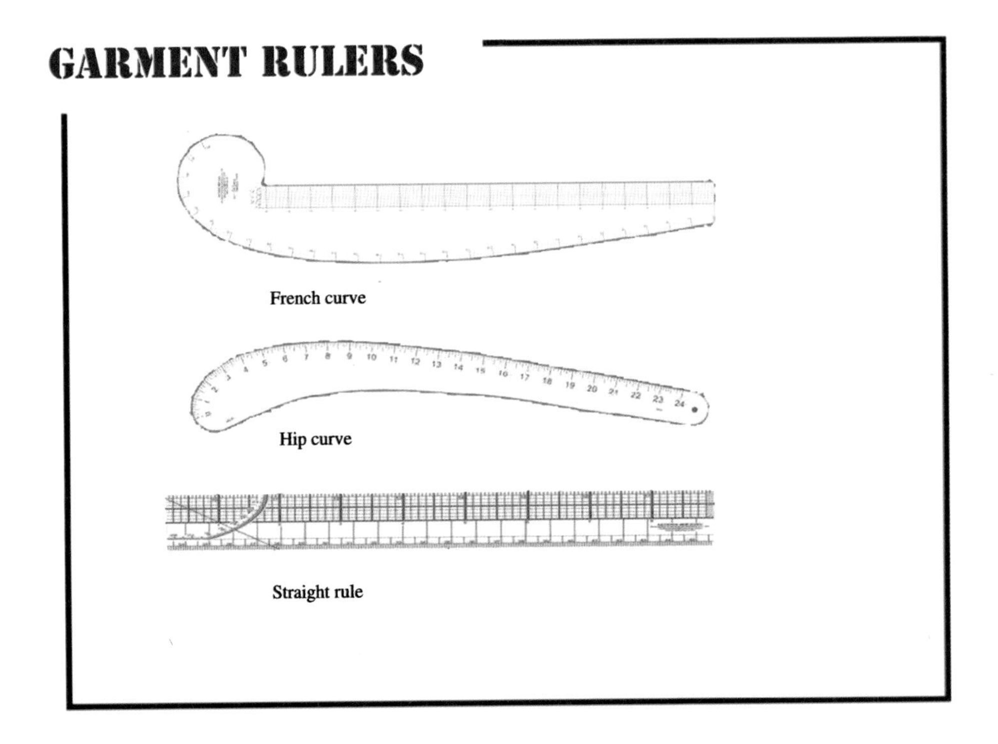

French curve

Hip curve

Straight rule

Words and expression

french curve 法式曲线板

hip curve 臀线曲线板

straight ruler 放码尺

Exercise Ⅱ:

According to the words and expression, describe the tools in your toolbox in English. 请根据已经学习的单词,用英语描述你工具箱中装有的工具。

Module Five —— Accessory 服装配饰

Section A: Speaking

The Runway Show

Lucas: Hi, Nina. Good morning!

Nina: Hi, Lucas. Good morning!

Lucas: What about style for the runway next month? Any good idea?

Nina: For the hairstyle and makeup?

Lucas: Yes, based on the theme of the fashion show, it's more suitable to have simple and clean makeup.

Nina: I agree with you. I think the whole style should focus on the eyebrow, to match with light brown eye shadow, and use nude lipstick, to brighten the eye.

Lucas: Is this makeup the trendy nude look? How about to match with high combed hairstyle?

Nina: Let me see. Maybe it's better to have long hair to the shoulder. If to do what you said, it looks too formal.

Lucas: Ok, it's all up to you, you know, we just need a general proposal. For the detailed plan, it's better to leave it to the professional company. Which company are you prefer?

Nina: I recommend Tony Guy styling studio. The stylists are well trained, and they are familiar with runway style.

Lucas: Ok, please ask them to work out a detail plan for us.

Nina: No problem.

Lucas: Thank you, see you then.

Nina: You're welcome. See you.

服 装 秀

卢卡斯:嗨,妮娜,早上好。

妮娜:嗨,卢卡斯,早上好。

卢卡斯:下个月的走秀风格定了吗?有没有什么好主意?

妮娜:在发型和化妆方面的?

卢卡斯:对的。基于服装秀的主题,造型方面更适合简单、干净的妆面。

妮娜:我同意你的看法。我觉得整个妆面应该强调眉毛,搭配浅棕色眼影,唇部采用裸妆唇膏,提亮眼部妆容。

卢卡斯:裸妆妆容是现在的流行趋势吗?搭配高的盘发发髻怎么样?

妮娜:让我想想,也许到肩膀的披发更适合。如果采用你所说的发型,有点太正式了。

卢卡斯:好的,你来决定。你知道的,我只是需要一个大致的提案。对于细节的计划,交给更专业的公司去做更好。你比较喜欢哪一间公司?

妮娜:我比较推荐托尼·盖造型工作室。里面的造型师都经过良好培训,他们对走秀造型比较熟悉。

卢卡斯:好的,那就让他们给我们做一个比较细节的方案。

妮娜:没问题。

卢卡斯:谢谢,到时候见。

妮娜:不客气。再见。

Section B: Reading

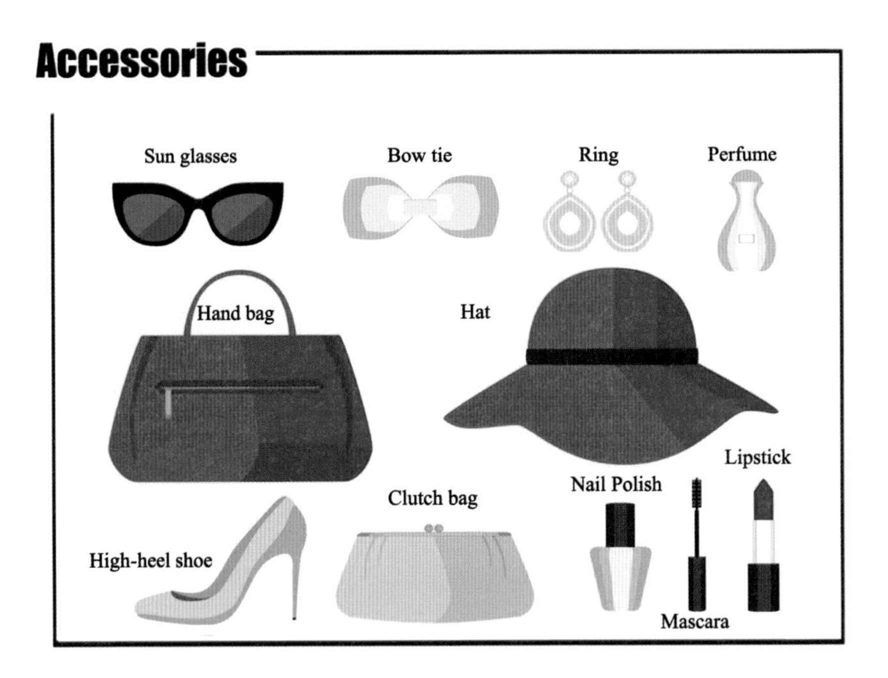

Words and expression

sun glass 太阳镜

bow tie 领结

ring 耳环

perfume 香水

hand bag 手提包

hat 帽子

high-heel shoe 高跟鞋

clutch bag 手拿包

nail polish 指甲油

mascara 睫毛膏

lipstick 口红

Words and expression

cosmetic 化妆品

hair styling 发型

shoes for ladies 女鞋

beauty & health 美容

accessories & clothes 服饰品

Exercise Ⅰ:

Write down the cosmetics you used frequently and your makeup steps. 写下你常用的几款化妆品和你的化妆步骤。

Exercise Ⅱ:

Collect the latest styling and makeup in European and American countries, Japan & Korea and China, and analyze the characteristics of them. 收集欧美、日韩和中国最近流行的时尚造型和妆容，分析各自的造型特色。

Project Three Marketing 服装营销

Module One —— Sales Ⅰ 服装销售Ⅰ

Section A: Speaking

Retail Skill of Clothing Ⅰ

Shelley: Hello. Can I help you?

Lady: Hi. I want to buy an overcoat. Can you help me to hold these two overcoats? I want to have a try.

Shelley: Sure. You can try them in the fitting room there.

Lady: How's that?

Shelley: It fits you so much. The shoulder, waist and length fit you. It shows your figure. This yellow makes you look younger.

Lady: I don't like the color, it's too light for me.

Shelley: It's ok. We still have dark blue and grey. Do you want to have a try as well?

Lady: Yes, I also want to try this pants.

Shelley: Ok, what size do you need?

Lady: Normally I wear size L.

Shelley: Ok, here you are.

(After trying on)

Shelley: Do you like these styles?

Lady: I like these styles, but not sure which one to choose. I feel a little hesitated.

Shelley: Now we have promotion. If to buy more than 500RMB, there will be discount 10% off.

Lady: Then can you work out how much would that be for these two?

Shelley: This coat and these pants cost 420RMB. Maybe you can buy one more accessories, like necklace. In this way, it can reach 500 RMB.

Lady: OK. I will choose a necklace. That's all now.

Shelley: AliPay or WeChat please?

Lady: AliPay.

Shelley: OK. Please scan the QR code. Welcome again.

Lady: Ok. See you.

Shelley: See you.

服装销售技巧Ⅰ

雪莉:你好,有什么需要帮忙的吗?

林迪:嗨,我想买一件外套。你能帮我拿一下这两件外套吗?我想要试试这两件。

雪莉:当然,你可以在那边的试衣间试一下。

林迪:你觉得我穿这件怎么样?

雪莉:很适合你啊,这个肩、这个腰线、这个衣长都非常适合你,很显身材啊。而且黄色把你衬得很年轻。

林迪:我不太喜欢这个颜色。对我来说太亮了。

雪莉:没关系,我们还有深蓝色和灰色。你想试一试吗?

林迪:好的,我还想试试这条裤子。

雪莉:好的,你穿什么码?

林迪:平时我穿L码。

雪莉:好的,给你。

(试穿之后)

雪莉:你喜欢这个款式吗?

林迪:我喜欢这个款式。但我不知道选哪一件,我有点犹豫。

雪莉:我们现在有促销。满 500 元人民币打九折。

林迪:那你帮我算一下,这两件多少钱?

雪莉:这件外套和这条裤子一共 420 元,也许你可以再买一些配饰,比如项链,这样的话就能达到 500 元了。

林迪:好的,我再选一条项链,完成任务。

雪莉:支付宝还是微信支付?

林迪:支付宝支付。

雪莉:好的,请扫这个二维码。欢迎下次再来。

林迪:好的,再见。

雪莉:再见。

Section B: Reading

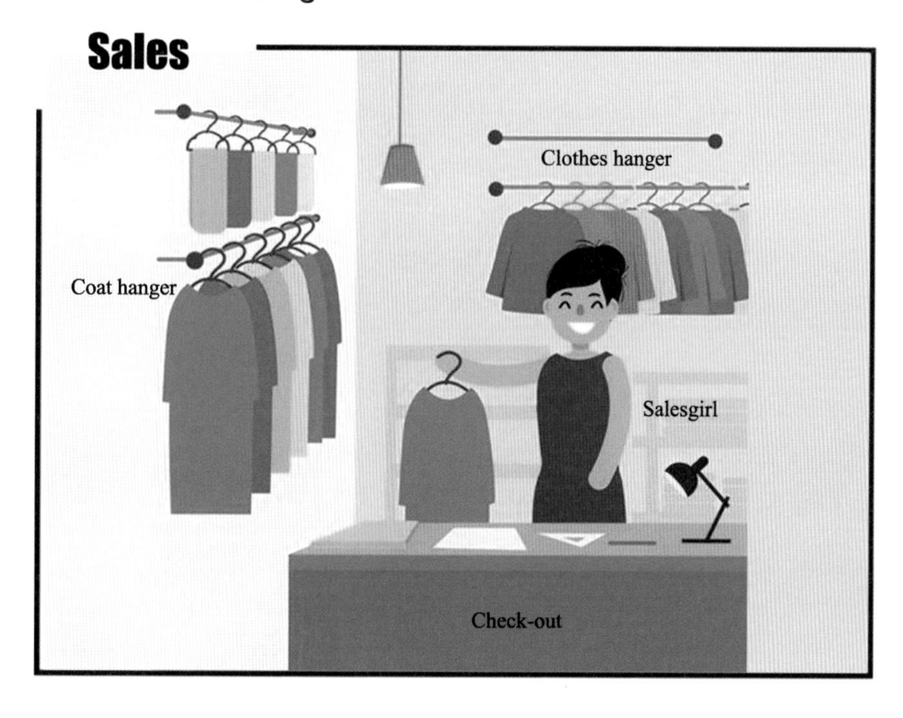

Sales

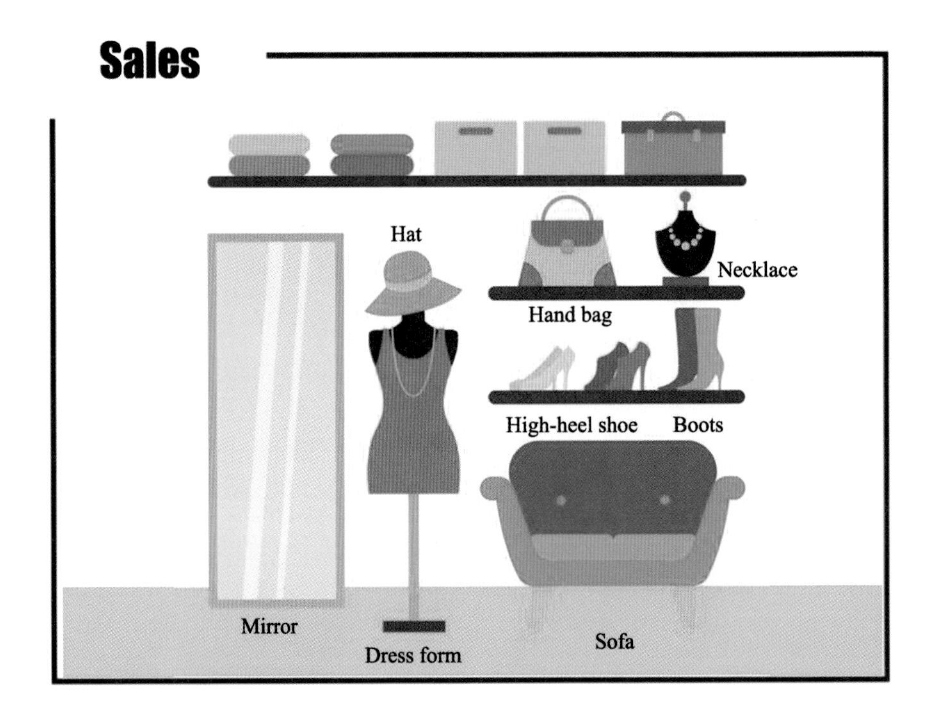

Words and expression

clothes hanger　服装挂杆

coat hanger　衣架

salesgirl　售货员

check-out　收银台

mirror　试衣镜

sofa　沙发

necklace　项链

boots　靴子

Exercise:

According to the words and expression, describe the experience of buying clothes in English. 请根据已经学习的单词,用英语描述一次你购买服装的经历。

Module Two —— Sales Ⅱ 服装销售 Ⅱ

Section A: Speaking

Retail Skill of Clothing Ⅱ

Jenny: Hi, welcome to ABC. Can I help you with anything today?

Lisa: Hi. Yes, actually. I'm looking to buy a present for my friend's daughter's birthday.

Jenny: I see, and how old is her daughter turning?

Lisa: She'll be 18.

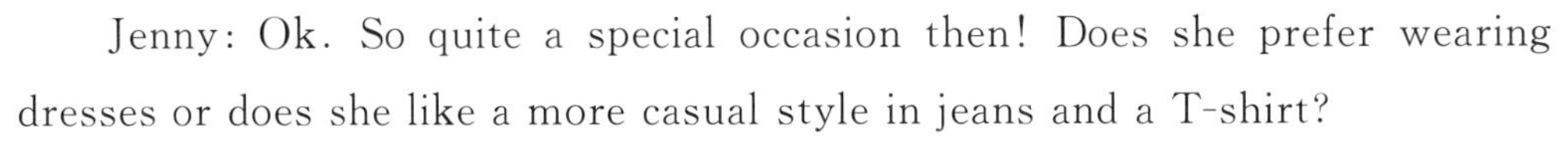

Jenny: Ok. So quite a special occasion then! Does she prefer wearing dresses or does she like a more casual style in jeans and a T-shirt?

Lisa: She's a jeans and T-shirt kind of girl, I think!

Jenny: Ok great, let's head over to our T-shirt range then. So here we have our new range of T-shirts, designed by Samantha.

Lisa: Perfect, thank you.

Jenny: So what sort of colors does she prefer? More neutrals or bold colors?

Lisa: I think she usually wears more neutral tone colors.

Jenny: Ok, so how about this grey cashmere top?

Lisa: Hmm, I'm not sure about those sleeves.

Jenny: Or how about this black and white striped T-shirt? This is one of our best sellers at the moment for young adults.

Lisa: Yes. That's nice, I like that. I can imagine her wearing that more than the gray one. I'll get one in a medium size please.

Jenny: Lovely, and this one is actually part of a 2-for-1 deal we have this weekend.

Lisa: Oh perfect! In that case, I'll also get this black T-shirt here for myself.

Jenny: Good choice, that works out well then! Would you also like some help picking out the jeans?

Lisa: Yes, please.

Jenny: Ok, follow me! So here we have our 5 main styles of jeans. Skinny jeans and straight leg for a tighter feel, or mom jeans, flares and boot cut for a more casual shape. I would say that skinny jeans are most popular amongst teenage girls.

Lisa: Skinny jeans it is!

Jenny: Great, how about this pair in the darker blue color?

Lisa: That does look nice. How much is that pair please?

Jenny: This pair is RMB499.

Lisa: Oh that's a bit over budget for me.

Jenny: Or how about this color here, this is a similar style but in a lighter wash, and these are RMB249?

Lisa: That's great. I'll go for that pair then.

Jenny: Perfect. Do you know her size in jeans?

Lisa: Yes, it's size 12.

Jenny: Ok. There you go, here's a size 12. And is that everything today? Can I help you with anything else?

Lisa: No. That's great, thank you.

Jenny: Great, if you'd like to follow me over to the till then and I'll put these through for you.

Lisa: OK. Thanks!

服装销售技巧Ⅱ

杰妮:你好,欢迎光临 ABC,有什么需要帮忙的?

丽莎:你好,有的。我在给朋友的女儿挑选生日礼物。

杰妮:明白。她女儿多大了呢?

丽莎:就要 18 了。

杰妮:那还是挺重要的。她喜欢穿裙子还是喜欢比较随意一点的 T 恤牛仔裤?

丽莎:我觉得她是个喜欢穿 T 恤牛仔裤的女孩。

杰妮:好的,那我们就去 T 恤区看一下。这边是我们新系列的 T 恤,是萨曼莎设计的。

丽莎:太好了,谢谢。

杰妮:她喜欢什么颜色?中性一点还是鲜亮一点?

丽莎:她平时穿的比较偏中性色。

杰妮:你觉得这个灰色的羊绒上衣怎么样?

丽莎:我不太喜欢这个袖子。

杰妮:那这个黑白条纹 T 恤呢?这个是目前年轻人买得最多的一款。

丽莎:这个挺好的,我喜欢。我能想象她穿这件好过那件灰色的。请给我一件 M 码。

杰妮:好的。这件其实是周末做活动买一送一的。

丽莎:那太好了,这样的话,我就再拿一件黑色的,我自己穿。

杰妮:选得好,这样就刚好了。你需要我帮你选牛仔裤吗?

丽莎:是的。

杰妮:那跟我来。我们主要有 5 种类型的牛仔裤。稍微修身一点的紧身裤和直腿裤。或者休闲一点的妈妈裤、喇叭裤和靴裤。我觉得青少年的女孩比较喜欢紧身裤。

丽莎:那就紧身裤吧。

杰妮:好的,这条深蓝色的怎么样?

丽莎:这条很好看,多少钱?

杰妮:这条 499 元人民币。

丽莎:这个有点超出我的预算了。

杰妮:那这个颜色呢? 款式差不多,就是颜色淡一些,这条 249 人民币。

丽莎:那太好了,我就买这条吧。

杰妮:好的,你知道她裤子穿多大码吗?

丽莎:知道,她穿 12 码。

杰妮:给你,这是 12 码的。今天就买这些吗? 有没有其他可以帮你的?

丽莎:没有了,谢谢。

杰妮:好的,那请跟我到柜台,给你结账。

丽莎:好的。谢谢。

Section B: Reading

Sales I
That's a bit steep. Can you do better on the price?
It's designer. Perfect quality. It's already a steal!
Can't you go any lower?
I'm sorry but that's the best I can do.

Sales I
We'll just take our money elsewhere...
Okay, how about $45? You won't find it cheaper than that anywhere else.
Done!

Words and expression

come in　有(货)

steep　过高的,过分的

designer　设计款

steal　很便宜

elsewhere　其他地方

anywhere　任何地方

done　成交

Exercise Ⅰ:

According to the words and expression, translate reading into Chinese.
请根据已经学习的单词,将阅读的英文部分翻译成中文。

Exercise Ⅱ:

According to the words and expression, describe the experience of buying shoes in English. 请根据已经学习的单词,用英语描述一次你购买鞋子的经历。

____

____

____

____

____

____

Module Three —— Sales Ⅲ 服装销售Ⅲ

Section A: Speaking

Retail Skill of Clothing Ⅲ

Coco: Hello. Can I help you?

Amber: Well, I am looking for some winter clothes for my fiancee.

Coco: Oh, it's the high time for you purchasing in our shop. We are now having a preseason sale on all our winter apparel.

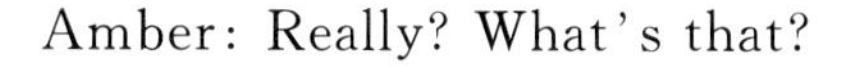

Amber: Really? What's that?

Coco: Everything for winter is 20% off.

Amber: I think my fiancee may favor the sweater in the shop window. Would you like to help me looking for any skirts that go with this sweater?

Coco: Sure, we have both skirts and trousers that would look well with the sweater. Look at this section.

Amber: I really like this flowery skirts. My fiancée will look very elegant in this.

Coco: You have a good taste. It's very much in style this year.

Amber: I don't think the green one fits her complexion. Do you have any skirts in light color or tan?

Coco: Look on the rack to the right.

Amber: Oh, yes. I'll take this one. Can I pay by WeChat?

Coco: It's ok.

服装销售技巧 Ⅲ

可可:你好,你想买什么?

安泊:你好,我想帮我未婚妻买一些冬天的衣服。

可可:哦,这真是在我们店购物的好时机。我们所有的冬季服装都在进行季前促销活动。

安泊:是吗?活动折扣如何?

可可:所有冬装八折优惠。

安泊:我感觉我未婚妻会喜欢你们橱窗里的那件毛衣。你能帮我挑一些和毛衣搭配的裙子吗?

可可:当然可以。我们有和毛衣非常搭配的裙子和裤子可选。请看看这个区域。

安泊:我很喜欢这件花裙子。我未婚妻穿上它应该会很优雅。

可可:你真有品味。这裙子今年很流行。

安泊:我觉得她的肤色不太适合这款绿色,你们有浅色或深色的裙子吗?

可可:你可以看看右边货架上的裙子。

安泊:哦,真好。我要买这件。我可以微信付款吗?

可可:可以的。

Section B: Reading

Sales II
Great! In that case, can you help me look for a dress to match my high heels?
Sure. This way, please.
50%

Sales II
I like this dress.
You have good taste. It's in style this year.
$32

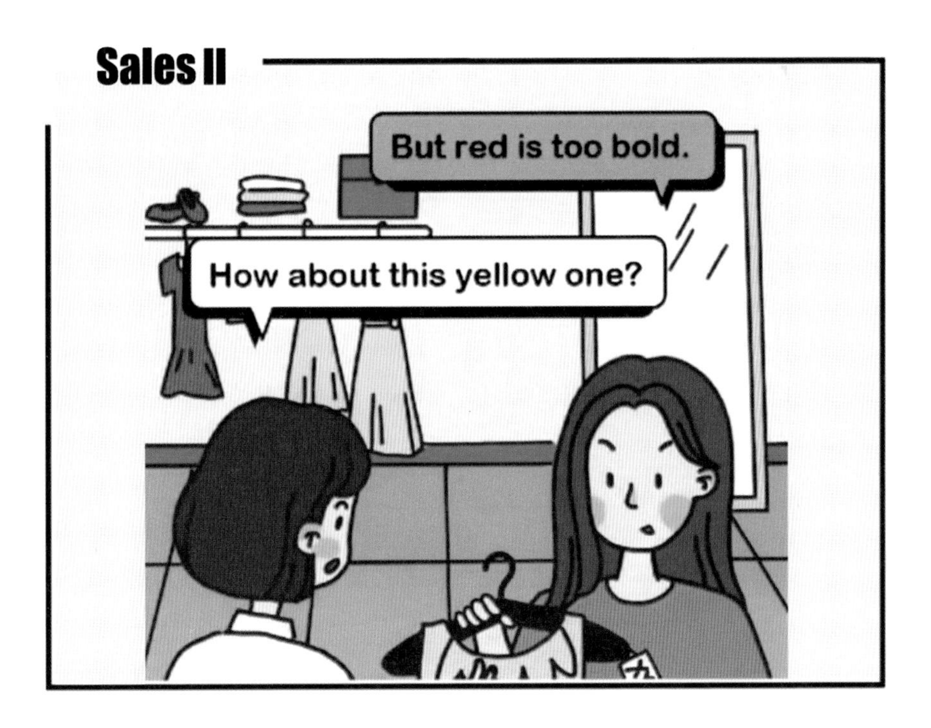
Sales II
But red is too bold.
How about this yellow one?

Sales II
That can work but it's too big. Do you have a smaller size?

Words and expression

preseason sale　季前促销

in that case　这样的话

look for　寻找

match　搭配

taste　品味

in style　流行款式

bold　显眼

Exercise Ⅰ:

According to the words and expression, translate reading into Chinese. 请根据已经学习的单词,将阅读的英文部分翻译成中文。

____

____

____

____

____

____

Exercise Ⅱ:

According to the words and expression, describe the experience of buying accessory in English. 请根据已经学习的单词,用英语描述一次你购买服饰品(配饰、箱包等)的经历。

____

____

____

____

____

____

Module Four —— After-sale service 售后服务

Section A: Speaking

Customer Service

David: Hello, customer service, this is David here. How can I help you?

Cindy: Hello, my name is Cindy. I'm calling because I'd like to return a handbag but I'm not sure how to do so.

David: That's ok, do you still have the box it came in?

Cindy: Yes, I do.

David: Excellent, then in that case put the handbag into the box, seal it up, and post it back to us.

Cindy: Do I have to pay for postage?

David: No, we will cover the postage. I'll send you an email with a prepaid address stamp on it. When you get it please print it out and attach it to the box then you can either arrange for a private delivery or post in the mail like normal.

Cindy: Great!

David: One more thing, could I have your order number please?

Cindy: Yes, of course. It's 123456789.

David: That's all set and your refund will be processed and will go through when the bag returns to us. Is there anything else you need assistance with?

Cindy: No, I'm good, thank you.

David: Ok, have a nice day.

Cindy: You too! Goodbye.

客　服

大卫:你好,客服中心,我是 David,有什么可以帮您?

辛迪:你好,我叫 Cindy,我想退一个包,但我不知该如何操作。

大卫:没关系,你是否还留有包装盒?

辛迪:是的,我有。

大卫:非常好,这样的话就把包放进盒子里然后封口,寄回给我们就好了。

辛迪:我需要承担寄回的运费吗?

大卫:不需要,我们会承担寄回的邮费。我会发送给你一封邮件,里面有我们已经付好邮费的寄回给我们的邮寄标签。当你收到后可以打印出来贴在箱子上,然后预约取件或者邮寄出来即可。

辛迪:太好了。

大卫:还有一件事,可以给我你的订单编号吗?

辛迪:当然,我的订单编号是 123456789.

大卫:好的,设置好了,你的订单将会为您退货退款。当我们收到包裹就会为您进行退款。还有什么需要我帮助的吗?

辛迪:没有了,谢谢。

大卫:好的,祝您生活愉快。

辛迪:你也是,再见。

Section B: Reading

After-Sale

- No. Too late. You can't give it back.
  不行，太迟了，你不能退货。
- I'm sorry, okay? I wasn't looking...and the store said they won't take it back because you signed for it.
  对不起，我没有注意，店里不肯接受退货，因为你已经签收了。

Words and expression

I was wondering…… 我想知道……

exchange 交换

pair 双

give it back 退货

sign 签名、签收

Exercise:

According to the words and expression, describe the experience of after-sale service in English. 请根据已经学习的单词,用英语描述一次你购买服装售后服务的经历。

____

____

____

____

____

____

Module Five —— Shopping 购物

Section A: Speaking

Shopping with your close friend

Bella: This one? Really?

Tiffany: Yeah. It's good.

Bella: So you don't think it's ugly?

Tiffany: No, I don't think so, it's cute.

Bella: Oh! The heat! Oh my god, today is so hot.

Tiffany: Yeah.

Lucas: Let's get some cold drink.

Tiffany: Emm. I'm good, I'm not thirsty.

Lucas: Really? Ok. I'm gonna have that one, thank you. Oh my god, cool.

Tiffany: Wow, that looks so good, can I have a try?

Lucas: Emm, of course.

Tiffany: Thank you. Emm, good.

Lucas: I told you.

Tiffany: Oh my god.

Lucas: I think it's just too oversized.

Tiffany: I told you, I think it's cute.

Lucas: Yeah, but I have a really strong arm.

Tiffany: Ah! You were right, it's so hot. Give me some of that. Here, I'm done. Oh my god, that looks so good, let's take a look.

Tiffany: Oh, Yeah.

和闺蜜一起逛街

贝拉:这件? 你说真的?

蒂芙尼:是啊。这件挺好的。

贝拉:所以,你不觉得这件丑吗?

蒂芙尼:不,我不觉得,挺可爱的。

贝拉:哦,好热,天啊,今天也太热了。

蒂芙尼:是的。

贝拉:那我们去买点冰饮吧。

蒂芙尼:嗯,不用了,我不太渴。

贝拉:真的吗,那好吧。我要那款冰饮,谢谢。哦,天呐,太爽了。

蒂芙尼:哦,看起来很好喝,我能尝尝吗?

贝拉:嗯,当然可以。

蒂芙尼:谢谢。嗯,真好喝。

贝拉:我和你说了吧。

蒂芙尼:哦,天呐,真好喝。

贝拉:我觉得刚刚那件衣服太大了。

蒂芙尼:我和你说了,挺可爱的。

贝拉:是的。但是我胳膊实在太粗了。

蒂芙尼:啊,你是对的。天真的太热了,再给我喝点你的冰饮吧。给你,我喝完了。哦,天呐,那个看起来超棒的,我们过去看看。

贝拉:哦,好吧。

Section B: Reading

Inquiry

- **这玩意儿打折吗?** Is there any discount?
- **能算我便宜一点吗?** Can you give me a little deal on this?
- **这家伙尺码正吗?** Is the measured size accurate?
- **这个有点太大了，我需要小一号。** It's too big for me, could you get a smaller size for me please?
- **这个有点紧。** It's a little tight.
- **小码/中码/大码/加大码** small/medium/large/extra large
- **这相机/衣服/电视/...有色差吗?** Is there any significant difference in color?
- **这邮寄得花多久?** How long does it take?
- **可不可以免邮啊亲?** Is free shipping available?
- **这个多少钱?** How much is this?

Inquiry

- **打X折：** XX% OFF, XX% Dsicount（比如8.5折就是15% OFF）
- **满减：**
  1. $100 OFF $500, 就是满$500立减$100（每满$500立减$100是$100 OFF Every $500）
  2. Save $100 on Every $500, 也是满$500立减$100的意思
- **额外XX折：** Extra XX% OFF（比如额外六折就是Extra 40% OFF）
- **包邮：** Free Shipping/ Shipping is Free
- **满XX免邮：** Free Shipping over $XX
- **可直邮：** International Shipping Available/ Shipping to countries outside U.S. is available
- **免退运费：** Free Return(s)

Inquiry

- **购物满XX即送礼包/礼卡/小样：** Gift/ Gift Card/ Samples with purchase of $XX
- **最高减$XX：** Up to $XX OFF
- **高达XX% OFF/低至XX% OFF:** Up to XX% OFF
- **超值套装：** Value Set
- **涨价/降价：** The price goes up/ down
- **清仓甩卖：** Clearance sale
- **购物袋/车：** Shopping bag/ cart
- **结账：** Check-out
- **愿望清单：** Wish list
- **我想买个XXX。** I'd like to buy a XXX/ I'm looking for a XXX
- **你能推荐一些给我吗？** Do you have any recommendations?

Words and expression

inquiry　询价

accurate　精确的

tight　紧的，紧身的

medium　中号的，中等的

significant　显著的

available　可获得的

sample　样品、样衣

value　价值

set　一套

clearance　清仓

recommendation　推荐、建议

Exercise:

According to the words and expression, describe an interesting conversation between you and your close friend in English. 请根据已经学习的单词,用英语描述一次你和你好朋友之间发生的有趣对话。

Reference(参考文献)

[1] 王强,白莉红,张爱华. 服装专业英语[M]. 2 版. 北京:化学工业出版社,2011.

[2] 柴丽芳,潘晓军. 服装实用英语[M]. 北京:中国纺织出版社,2012.

[3] 严国英,庄福珍. 服装专业英语[M]. 2 版. 北京:中国纺织出版社,2003.

[4] Sophie. Shopping in the mall, choosing clothes, situational dialogue. [EB/OL] [2021-12-18]. https://www.xiaohongshu.com/discovery/item/5ef5f4eb000000000101c100?xhsshare=CopyLink&appuid=555aad989a5fe334fbae7b94&apptime=1641109921.

[5] Karina Hernandez. Fashion Product Development Job Description [EB/OL] [2021-09-21]. http://www.ehow.co.uk/zbout_6284385_fashion-product-development-job-description.html.

[6] Deepak Singh. Pink is out but blue is in. Understanding Fashion & trend forecasting [EB/OL] [2022-01-03]. http://www.fibre2fashion.com/industry-article/9/836/pink-is-out-but-blue-is-in-understanding-fashion-trend-forecasting1.asp.

[7] Bean By The Window. English words. [EB/OL] [2022-01-11]. https://www.xiaohongshu.com/discovery/item/60e5519a000000000102ffd0?xhsshare=CopyLink&appuid=555aad989a5fe334fbae7b94&apptime=1641115949.